THE HACKING

of the

AMERICAN MIND

THE HACKING

of the

AMERICAN MIND

The Science Behind the Corporate
Takeover of Our Bodies and Brains

Robert H. Lustig, MD, MSL

AVERY • AN IMPRINT OF PENGUIN RANDOM HOUSE • NEW YORK

AVERY

an imprint of Penguin Random House LLC
375 Hudson Street
New York, New York 10014

Most Avery books are available at special quantity discounts for bulk purchase for sales promotions,
premiums, fund-raising, and educational needs. Special books or book excerpts also can be created
to fit specific needs. For details, write SpecialMarkets@penguinrandomhouse.com.

Library of Congress Cataloging-in-Publication Data

Names: Lustig, Robert H., author.
Title: The hacking of the American mind : inside the sugar-coated plot to
confuse pleasure with happiness / Robert H. Lustig, M.D., M.S.L.
Description: New York : Avery, 2017. | Includes bibliographical references and index.
Identifiers: LCCN 2017031139| ISBN 9781101982587 (hardcover) | ISBN 9781101982594 (epub)
Subjects: LCSH: Happiness. | Pleasure. | Contentment. | Satisfaction.
Classification: LCC BF575.H27 L83 2017 | DDC 152.4/2—dc23
LC record available at https://lccn.loc.gov/2017031139

Printed in the United States of America
1 3 5 7 9 10 8 6 4 2

Book design by Meighan Cavanaugh

Dedicated to my late mother, Judith Lustig Jenner (1934–2016), the inspiration for this book. My mother wasn't a particularly happy person. A Depression baby, she had to grow up quickly, and was an adult by the age of four. She missed out on a real childhood, and spent the rest of her life trying to make up for it. To her, money was the route to happiness, and she didn't want for it, but it never really made her happy. She certainly knew pleasures—in food and drink, in jewelry, in casinos, in exotic spots around the world. But few of her exploits or possessions brought her contentment. The only true happiness she knew were her children and grandchildren, and her eight-year relationship with her second husband, Myron Jenner, who was taken all too soon. Along the way and at the end, she also knew a large dose of pain and suffering as her body broke down from a debilitating neurological illness while her mind stayed as sharp as a tack. Rest in peace, Mom. I have no doubt that the happiness that eluded you in this world will be yours in the next.

CONTENTS

INTRODUCTION

Happiness is neither virtue nor pleasure nor this thing nor that, but simply growth. We are happy when we are growing.

—John Butler Yeats to his son William Butler Yeats, 1909

We were all children once. Like you, more chance than not, my greatest moments of happiness during childhood have stuck with me, and to this day continue to bring a smile, and sometimes even a tear. Childhood is a time of mind expansion—not just in knowledge but in experimentation, in inquisitiveness, in trying out new concepts and strategies. Childhood is supposed to be a time when the balloon of happiness soars high above the mundane. The tools of the trade for most kids were a peanut butter sandwich, a bicycle, and a bedtime story. I became a pediatrician, in part, to relive and help channel the wonder and delight involved in growth.

Fast-forward four decades. Children still grow, but sadly in my pediatric clinic I now watch many of them grow horizontally rather than vertically. Some take medicines previously reserved for adults, like metformin for type 2 diabetes or benazepril for hypertension. And that balloon of happiness, that sheer wonder of it all, is now so deflated, there isn't enough buoyancy for it to soar. Rather, in its place has been dropped some weighty pleasures of the mundane, in this thing or that. Standard issue now are Capri Sun, Netflix, and Snapchat.

You might argue, well, that's progress, that's convenience, that's technology, that's our new instant gratification culture—buy a pleasure to increase happiness. But what if those pleasures, ostensibly developed and marketed in the name of increasing your happiness, actually did the opposite? What if they actually made you unhappy? What if they changed your brain so that happiness was sapped from you? What if today's kids are actually canaries in the coal mine? What if these same brain changes extended to your coworkers, to your friends, to your family members, and to you? For better or worse? And better for whom?

Pleasure and happiness are similar, as they both feel good. But Yeats knew they weren't the same. Since the recording of time, philosophers have tried to wrestle these two positive emotions to ground. These two uniquely human phenomena have together and separately occupied outsized parcels of our consciousness, our literature, and our national and international discourse. While our philosophers and social commentators have spent the last three thousand years defining and redefining these two terms for us, something quite unusual and likely even sinister has befallen these related yet decidedly different positive emotions.

These past forty years have witnessed the twin epidemics of the negative extremes of both of these emotions: addiction (from too much pleasure) and depression (from not enough happiness). Yet in these same forty years our knowledge of brain science has advanced to the point where these two emotions can now be dissected and parsed at a biochemical level. Did the uptick in prevalence of addiction and depression occur naturally? Separately? In a vacuum? Or under some form of outside pressure? What, or who, has ushered modern society into this new normal? What if all of Western society has been hacked, to profit a few at the expense of the many? And what if you didn't even know you'd been hacked?

"Hack" is a word with a relatively short history in our modern lexicon, with a fluid meaning. The first reference to a "hack" was at a meeting of the MIT (my alma mater) Model Railroad Club in 1955. At that time "hack" meant a "prank" whose perpetrators demonstrated style, resourcefulness, and whimsy in its performance. Stealing a car is a felony offense. Stealing a Boston Police Department vehicle, disassembling it, carrying each piece up five floors, and then reassembling it at the top of the Great Dome at MIT, complete with a life-size policeman mannequin and a box of doughnuts in the front seat—now *that* is a *hack*. More recently Silicon Valley types stole the word to denote clever solutions to difficult problems, known as "white hat" hacking. Yet "black hat" hacking dates back to 1963, when an unauthorized hacker remotely commandeered the MIT mainframe computer. As computers became more interconnected and more technologically advanced, less whimsical people started to create viruses to infect other computers, and hacking took on a much more ominous and sinister tone. As we all learned from the 2016 election debacle, today's computer hacking encompasses three steps. Step one is pfishing, where a seemingly benign yet imperative e-mail message with a disguised zipfile or URL is sent to an unsuspecting victim; if the message is clicked, that computer is rendered vulnerable and the hacker can gain entry. Step two is the insertion of some form of malicious code into the victim's computer. Depending on the goals of the hacker, step three is the hijacking of something—for instance, the material stored in a computer's memory (like Democratic National Committee e-mails), which is transferred to the hacker, who can use it to humiliate or blackmail; or the computer's executable files, in order to hold the computer for ransom; or even the victim's hard drive, which can be crashed and erased, the ultimate in malevolence.

You say, well, that's computers . . . What does this have to do with the human body or brain? How about everything? While human hacking does not occur via computer code, there are many ways to tinker with the human brain. Certainly drugs can do the tampering. How about cleverly disguised messages, disinformation, propaganda, and the newest method of tampering, fake news? Can these messages act like phishing? And what if one of these messages gains hold? Can these alter your brain? Or how about something as innocuous as food? All of the above.

In this book I am going to develop separate and parallel scientific, cultural, historical, economic, and social arguments that our minds have been hacked. I will also demonstrate that this hack—the systematic confusion and conflation of the concepts and definitions of pleasure and happiness—has been inserted into the limbic system (the emotional part) of our brains, thereby precipitating a slow-motion crash of a substantial percentage (somewhere between 25 and 50 percent) of individuals and exacting a severe detrimental impact on our whole society. I will also demonstrate that this hack wasn't accidental but in fact has been a plot— that is, the hack was not to just create mischief; rather, it was specifically designed and engineered with a profit motive. And, similar to the Russian hack of the 2016 presidential election, this plot has been and continues to be executed by private interests with governmental support.

In order to convince the reader of each of these arguments, I will first lay out (in simple terms) the neuroscience of each of these two otherwise positive emotions, how they can sometimes appear similar, but more importantly how they differ, what underlies our experience of each one, and how they influence each other. I will then explain how the business community and government have taken advantage of this neuroscience to hack our decision-making capacity and alter our level of individual and collective well-being. But fear not: even though this plot is pervasive in all walks of life, there are ways to insulate yourself and fend off this hack. Because when we understand the neuroscience of pleasure and happiness, each one's relationship to the other, and how they are manipulated by our current food, technology, and media environments, we can more accurately denote the causes—and in turn the treatments—for our own personal well-being, and for our twin societal scourges.

I am not a psychiatrist or an addiction specialist. I am not a motivational speaker or a pop culture icon. I am not a Buddhist or a self-help guru. I am definitely not Dr. Oz or Dr. Phil, nor do I want to be: those guys have got their own problems. I am not a purveyor or user of psychoactive substances (although I've consulted with some experts for this book). I am not even a strict practitioner of all the precepts elaborated in this volume. And I certainly don't have a corner on either the pleasure or happiness markets. Hell, I've got my own issues and baggage.

I am a practicing pediatric endocrinologist (hormone problems in children) and obesity research scientist at UCSF, an academic medical center. Endocrinology is a profession that has morphed over the past three decades from one that previously generated great joy and satisfaction into one of the unhappiest occupations around. Burnout rates are at 54 percent of all doctors but 75 percent of endocrinologists. Our subspecialty takes care of patients with obesity who never get thinner and patients with diabetes who never get better, most of whom eschew the advice that we recommend and destroy their bodies and their minds in the process. The practice of endocrinology is particularly prone to mythology and quackery, because hormones are chemicals you can't see. People can see the damage that smoking does to their lungs on X-rays, or to their hearts on catheterization. But you can't see the hormones at work in obesity and diabetes. And so people don't believe. For many people, *not seeing is believing*. And charlatans can make people not see anything they want.

I'm not a conspiracy theorist by nature. A conspiracy would suggest corporate malevolence with collusion between industry actors, with intended malice and with government approval. Woodward and Bernstein had to connect many dots before the pernicious nature and the smoking gun of Watergate was revealed. It took whistleblower Jeffrey Wigand and the publication of the "tobacco documents" before officials could demonstrate that tobacco industry executives were engaged in a conspiracy to defraud the public. In *The Hacking of the American Mind*, I have a lot of dots that I must connect for you in successive chapters (biochemistry, neuroscience, genetics, physiology, medicine, nutrition, psychology/psychiatry, public health, economics, philosophy, theology, history, law). Although there are indications that some of the perpetrators (like tobacco) have colluded, or at least shared data and practices, I'm going to declare right now there is no smoking gun (other than smoking), and so I'm not going to stick my neck so far out as to say that there has been a conspiracy between different industries and the government to purposefully inflict malice on the public. Nonetheless, I will argue that there has been a plot by some industries to obfuscate the link between their products and disease, and to willfully confuse the concepts of pleasure and happiness with the sole motive being profit. I will then tie these seemingly separate strands together to convince the reader of the new alt-reality that has been manufactured by these industries. The science, the history, and the politics are strong enough to provide circumstantial and empirical evidence. In successive parts of this book, I will elaborate on each of these.

The substance that got me started on thinking about nutrition, health, disease, and how our emotions are manipulated—the substance that revealed its hidden iniquities to me back in 2006—is sugar. Sugar is the other white powder. It was the science of sugar that showed me that the behaviors associated with obesity (gluttony and sloth) were in fact due to a change in biochemistry, and that the biochemistry was due to a change in the environment. You may have read my book *Fat Chance*, which asked two questions: Why are we all so fat and sick? And in just thirty years? *Fat Chance* is a treatise on the science of obesity and metabolic syndrome, and the implications that the science portends for people and policy. But it was understanding the brain science that allowed me to put the data together to form a unifying hypothesis, and that sparked the impetus for my efforts to educate the public—to debunk the myths surrounding the obesity epidemic, which had prevented policy makers from addressing the deficiencies of our toxic food environment, rather than ineffectively trying to modulate the behaviors that are the result of that biochemistry. This meant I needed to know the law surrounding public health in order to understand and impact policy. So in my sixth decade I went to law school.

In the process of putting together the scientific argument in *Fat Chance* for nutrition and *physical* health, it became apparent to me that there is a wealth of information on the role of nutrition on outcomes related to *behavioral* health. Yet this information remains virtually unknown to most doctors and patients. Worse still, entire industries and governments have pushed hedonic (reward-generating) substances and behaviors on their unsuspecting populations for their profit, which has only caused further unhappiness. I also came to realize that some of the basic tenets of modern medicine were simply rubbish. They may sound right, but they do not stand up to scientific scrutiny.

This book, *The Hacking of the American Mind*, is similar to *Fat Chance* in that it uses biochemistry to educate the reader about the toxic environment in which we currently find ourselves—and perhaps even more importantly, how we remain there. (As was true in *Fat Chance*, the punch line is that it's not about personal responsibility, but only you can help yourself, because no one else will.) Because pleasure and happiness, for all their apparent similarity, are separate phenomena, and in their extreme function as opposites. In fact, pleasure is the slippery slope to tolerance and addiction, while happiness is the key to long life. But if we don't understand what's actually happening to our brains, we become prey to industries that capitalize on our addictions in the name of selling happiness.

At this point it's essential to define and clarify what I mean by these two words—*pleasure* and *happiness*—which can mean different things to different people.

Merriam-Webster's Dictionary defines "pleasure" as "enjoyment or satisfaction derived from what is to one's liking"; or "gratification"; or "reward." While "pleasure" has a multitude of synonyms, it is this phenomenon of *reward* that we will explore, as scientists have elaborated a specific "reward pathway" in the brain, and we now understand the neuroscience of its regulation. Conversely, "happiness" is defined as "the quality or state of being happy"; or "joy"; or "contentment." While there are many synonyms for "happiness," it is the phenomenon that Aristotle originally referred to as *eudemonia*, or the internal experience of contentment, that we will parse in this book. Contentment is the lowest baseline level of happiness, the state in which it's not necessary to seek more. In the movie *Lovers and Other Strangers* (1970), middle-aged married couple Beatrice Arthur and Richard Castellano were asked the question "Are you happy?"—to which they responded, *"Happy? Who's happy? We're content."* Scientists now understand that there is a specific "contentment pathway" that is completely separate from the pleasure or reward pathway in the brain and under completely different regulation. Pleasure (reward) is the emotional state where your brain says, *This feels good—I want more*, while happiness (contentment) is the emotional state where your brain says, *This feels good—I don't want or need any more.*

Reward and contentment are both positive emotions, highly valued by humans, and both reasons for initiative and personal betterment. It's hard to be happy if you derive no pleasure for your efforts—but this is exactly what is seen in the various forms of addiction. Conversely, if you are perennially discontent, as is so often seen in patients with clinical depression, you may lose the impetus to better your social position in life, and it's virtually impossible to derive reward for your efforts. Reward and contentment rely on the presence of the other. Nonetheless, they are decidedly different phenomena. Yet both have been slowly and mysteriously vanishing from our global ethos as the prevalence of addiction and depression continues to climb.

Drumroll . . . without further ado, behold the seven differences between reward and contentment:

1. Reward is short-lived (about an hour, like a good meal). Get it, experience it, and get over it. Why do you think you can't remember what you ate for dinner yesterday? Conversely, contentment lasts much longer (weeks to months to years). It's what happens when you have a working marriage or watch your teenager graduate from high school. And if you experience contentment from a sense of achievement or purpose, the chances are that you will feel it for a long time to come, perhaps even the rest of your life.

2. Reward is visceral in terms of excitement (e.g., a casino, a football game, or a strip club). It activates the body's fight-or-flight system, which causes blood pressure and heart rate to go up. Conversely, contentment is ethereal and calming (e.g., listening to soothing music or watching the waves of the ocean). It makes your heart rate slow and your blood pressure decline.

3. Reward can be achieved with different substances (e.g., heroin, nicotine, cocaine, caffeine, alcohol, and of course sugar). Each stimulates the reward center of the brain. Some are legal, some are not. Conversely, contentment is not achievable with substance use. Rather, contentment is usually achieved with deeds (like graduating from college or having a child who can navigate his or her own path in life).

4. Reward occurs with the process of *taking* (like from a casino). Gambling is definitely a high: when you win, it is fundamentally rewarding, both viscerally and economically. But go back to the same table the next day. Maybe you'll feel a jolt of excitement to

try again. But there's no glow, no lasting feeling from the night before. Or go buy a nice dress at Macy's. Then try it on again a month later. Does it generate the same enthusiasm? Conversely, contentment is often generated through *giving* (like giving money to a charity, or giving your time to your child, or devoting time and energy to a worthwhile project).

5. Reward is yours and yours alone. Your sense of reward does not immediately impact anyone else. Conversely, your contentment, or lack of it, often impacts other people directly and can impact society at large. Those who are extremely unhappy (the Columbine shooters) can take their unhappiness out on others. It should be said at this point that pleasure and happiness are by no means mutually exclusive. A dinner at the Bay Area Michelin three-star restaurant the French Laundry can likely generate simultaneous pleasure for you from the stellar food and wine but can also generate contentment from the shared experience with spouse, family, or friends, and then possibly a bit of unhappiness when the bill arrives.

6. Reward when unchecked can lead us into misery, like addiction. Too much substance use (food, drugs, nicotine, alcohol) or compulsive behaviors (gambling, shopping, surfing the internet, sex) will overload the reward pathway and lead not just to dejection, destitution, and disease but not uncommonly death as well. Conversely, walking in the woods or playing with your grandchildren or pets (as long as you don't have to clean up after them) could bring contentment and keep you from being miserable in the first place.

7. Last and most important, reward is driven by dopamine, and contentment by serotonin. Each is a neurotransmitter—a biochemical manufactured in the brain that drives feelings and emotions—but the two couldn't be more different. Although dopamine and sero-

tonin drive separate brain processes, it is where they overlap and how they influence each other that generates the action in this story. Two separate chemicals, two separate brain pathways, two separate regulatory schemes, and two separate physiological and psychological outcomes. How and where these two chemicals work, and how they work either in concert or in opposition to each other, is the holy grail in the ultimate quest for *both* pleasure *and* happiness.

The Hacking of the American Mind will not just elaborate how reward and contentment work on a biochemical level, it will show what the differences between them mean for your personal and mental health and for the health of our society. However, right at the start, I must acknowledge three caveats.

First, the science of these two phenomena relies primarily on animal models. Who says depression in a rat is the same as depression in a human? Or even addiction, for that matter? Can rats become sex addicts?

Second, most human studies that are available are correlative, not causative. Correlation is a snapshot in time. You can only say that two things are related to each other. And even that can be a stretch. Might they have nothing to do with each other? For instance, ice cream consumption correlates with frequency of drownings. Does that mean eating ice cream causes you to drown? Or do survivors of the drowned victim bury their sorrows in a banana split? More likely, we eat ice cream when it's hot, we swim when it's hot, and some unfortunate people drown when they swim. Just because there is a correlation, does that really mean there is a cause-and-effect relationship?

There are other complications in interpreting human studies:

• It's very hard to do causative studies on emotions and psychiatric illness. Determining causation means assessing the disease process over time. Few people have had magnetic resonance imaging (MRI) or positron emission tomography (PET) scans of their brains performed before their mental illness occurred.

- Many of the studies measure blood levels of these neurotransmitters. However, what is going on in the brain may be different than what is going on in the blood.

- Brain neuroimaging studies require special equipment; some involve radioisotopes and are therefore terribly expensive to perform and often not immediately available.

- It's not just all dopamine and serotonin. Other neurochemicals do play major roles in how we think and feel, are part of these pleasure and happiness pathways as well, and thus complicate the picture.

- All of these pathways and neurochemicals are influenced by genetic, epigenetic (changes to the expression of DNA, not changes to the sequence), and experiential forces. Thus, what might be true for one individual may not be true for another.

- The science on serotonin was stymied for forty years by Congress and the FDA. I'll expand on this later in the book. But it means we have way less information on the role of serotonin on behavior than we should.

Third and finally, the connecting of our moods and emotions to rational public policy is complex, nuanced, and indirect. People can't be *told* what to do. As a New Yorker, I admit that if someone tells me to jump, my first response is not "How high?" But to have even a remote chance to unhack our brains, first we have to recognize what the hack is and how it works.

Part I will discuss the differences between reward and contentment, how their meanings have been confused and obscured, and how they indeed can be opposites. We will also start to explore what parts of the brain are involved in each experience. Part II will elaborate on the biology of reward and the science of dopamine. I will explain why the motivation for pleasurable experiences starts with dopamine but how too much of it can lead to aggression and irritability. There really can be too much of a good thing. It can even kill you. Throw on top of that some emotional stress, which aggravates the need for pleasure seeking, and you've got a great recipe for addiction. Part III will discuss the biology of contentment and the science of serotonin and how the reward and contentment systems overlap (or don't). For instance, certain serotonin agonists (like psychedelics) can improve mood, while other serotonin-boosting medications (known as selective serotonin reuptake inhibitors, or SSRIs) treat depression. In Part IV, I will show how the perpetration of this "plot" has brought us to this place, from a personal, historical, cultural, and economic standpoint. In the last half century, America and most of the Western world have become more and more unhappy, sicker, and broke as well. Marketing, media, and technology have capitalized on subverting our brain physiology to their advantage in order to veer us away from the pursuit of happiness to the pursuit of pleasure, which for them of course equals the pursuit of profit. Fueling our quest for reward has only contributed to the epidemics of non-communicable diseases such as diabetes, heart disease, cancer, and dementia, which are eating away at our health, our health care system, and the fabric of our society. Lastly in Part V, I will offer simple solutions that all of us can employ to defend against the pernicious peddling of pleasure, and ways to mitigate the stress that drives both addiction and depression, so that we may be able to pursue our individual happiness to the fullest. I will explore how and why different modalities for taming dopamine and increasing serotonin work and how we can rethink our lives and our goals so we can enjoy health (more than we have now) as well as pleasure (sometimes) and happiness (all the time). But you can't solve the problem until you know what the problem is. That's what this book is about.

Humans speak many languages, have varying standards of beauty, and worship at the altars of different deities, but their underlying bio-chemistry and what makes them tick is nonetheless the same. All our behaviors are manifestations of the biochemistry that drives them. To pull ourselves and our children back from the edge of this man-made abyss at which we now stand, we first have to understand the science.

PART I

A Few Fries Short of a Happy Meal

1.

The Garden of
Earthly Delights

Once upon a time we were happy. Then the snake showed up. And we've been miserable ever since. Hieronymus Bosch's painting *Garden of Earthly Delights* (circa 1500) is a triptych housed in the Prado in Madrid. It is an allegorical warning of what happens when we squander our birthright of happiness divined from God in one garden and move on to the pleasures of the flesh in the next garden, with the inevitable result of eternal damnation. Figures. Our most lauded goal in life—to be happy—is seemingly an illusion, out of reach for us common folk. Except the rich aren't any happier. Happiness seems to be a mirage, something to chase after, to keep us turning over rocks, kissing frogs, and trying to fit keys into the magic lock.

But along the way, wandering through our own individual gardens of earthly delights in search of our seemingly unobtainable nirvanas, we've sure had a whole lot of fun. Or we've at least tried to. We buy shiny things, play Powerball, imbibe with friends or sometimes alone. So why are so many of us miserable? Are we destined just to sink further into the abyss of pleasure with no hope of extricating ourselves to find real happiness? Is it all futile? Lots of people have died trying to get to that magic place of contentment and inner peace, that thing called "happiness." But if we can't get there, what's the point?

What if I told you that happiness is right there in front of you, just behind the curtain of your own brain?

To some, an argument over the difference between pleasure and happiness might seem like a straw man, a false argument not really worth having. Hey, they both feel good; why should you care? And pleasure is here, now. Happiness . . . maybe not so much, and not so soon.

But it does matter. And not just to you but to all of society. Explaining the differences between these two otherwise positive emotions forms the narrative arc of this book.

Terms of Endearment

Pleasure takes many forms and has many synonyms: "gratification," "amusement," "indulgence," "titillation," "turn-on." But the experience of pleasure is the visceral readout of activity of a specific brain area known as the "reward pathway." In fact, pleasure is actually two phenomena in one. First, one experiences the motivation for a given reward. Second, one experiences the consummation of that reward as the visceral experience we call pleasure. For simplicity, I will call it reward so both the social science and the neuroscience can effectively be treated as one.

The old adage goes, "Beauty is in the eye of the beholder." Same for happiness. Happiness is in the brain of the experiencer. And it too has its own brain area, known as the "contentment pathway." But as a philosophical concept, happiness has a long history and has been tangled up with the history of society for as long as there's been society. Happiness consists of a grab bag of definitions that have changed and morphed over time.[1] The root of the word, "hap," means luck. And we see this etymological root in other words relating to chance occurrence: for instance, *hap*penstance or per*hap*s. Early societies weren't very happy; after all, with famine, plague, and war, they had a lot to be unhappy about. Happiness was chance, fleeting, and seemed to alight on only a select few in any given society.

The God Factor

Religion has been the arbiter of both pleasure and happiness since there *was* religion. By no means is the brief history that follows meant to be exhaustive, but understanding where we came from can help us determine where we are going.

The Jewish tradition says that the study of the Torah is the path to happiness, because "all its paths are peace," and by following the law one could not help but achieve happiness. The Greeks are on record for jump-starting both the pleasure and happiness industries. In the third century BCE they wrestled the concept of happiness away from the concept of hedonism, the philosophy that said that the goal of life was net pleasure (pleasure minus pain). Aristotle expanded on the Jewish concept and argued that happiness consisted of being a good ethical person, a manifestation of reason and virtue, and coined the term *eudemonia*, a synonym for "contentment" (the concept on which this book is based). Zeno, the father of Stoicism, took this up a notch to say that unhappiness resulted from errors of judgment and that the true sage was immune to unhappiness; the converse of this was, of course, that if you were unhappy, you were no sage. Epicurus weighed in to say that happiness was a state of peace, absence of fear, absence of pain, and a life surrounded by friends—threads of which remain with us today.

Then came Christianity, which said many things, one of which was that happiness will occur there and later as opposed to here and now. Life is unpleasant, but if you live it as an upstanding Christian, heaven awaits. Pleasure was the devil on earth, and pain in the form of humility and service was the path to a happy afterlife, a gift from God. Islam refined the concept to turning it into a struggle, the war between good and evil on earth, and one would be rewarded with happiness in the afterlife. And the Baha'i faith has its feet in both camps by stating that we humans are noble from the start and capable of continual spiritual growth both in this world *and* in the afterlife. So make the world a better place now and heaven a better place later.

The Eastern religions take a slightly different approach, by establishing the methods for achieving happiness now rather than later, because there *is* no later—at least, not the heaven of Western theology. Hinduism proffered the theory of reincarnation as a means of "getting it right"—that the goal of religion was to adhere to a way of stopping the process of death-rebirth (so you don't come back as a frog). Buddhism added specific practices allowing us to break free of this cycle to achieve "nirvana," or liberation. Thus, pleasure has historically been the cultural antagonist to achieving happiness. In terms of the science, nothing's changed.

Indeed, there is not one definition of "happiness." What it means to be happy is quite different, depending on the times in which you live, your religious and cultural affiliations, and likely the language you use. For instance, some languages define "happiness" as "good luck and favorable circumstances" (i.e., out of your control), while in others "happiness" refers to "favorable internal feeling states" (somewhat in your control).[2] Obviously, this makes it very hard to write about, because the definitions and the criteria for inclusion have been a moving target.

Happy Endings?

Happiness is what most people say they really want: the spouse who can manage those things you can't; the house with the porch and the white picket fence; the two matched children (one boy, one girl) who get all the awards in high school and go on to Ivy League colleges; seeing the world with your family; having a retirement nest egg (I always liked the Prudential commercial with psychologist Dan Gilbert that states, "Retirement is paying yourself for what you like to do"); and growing old with your spouse without infirmity. Then again, most parents today simply wish for minimal psychiatric bills, no trips to rehab and no police record, good colleges on their children's résumés, and offspring who are neither bullies nor bullied. Yet virtually any hallmarks of happiness are noticeably absent from most of our written history,[3] in part because who'd want to read it? That's kind of the point. Happiness is what we *say* we want. But reading about someone else's happiness can get kind of boring. Lack of conflict doesn't make for a very good page-turner or miniseries.

Since the Renaissance, happiness has been the main stated goal of life, rather than being on good behavior to reserve yourself a seat in the afterlife. When asked their primary desire, people across the world, from the U.S. to Slovenia, have put happiness at the very top of their lists.[4] But despite our five-hundred-year eyes on the prize, as a whole we consistently miss the target. The self-help section of any bookstore (that is, any bookstore that is left: their disappearance is itself a marker of our collective loss of happiness) is chock-full of tomes that explore the achievement, value, or consequences of pleasure or happiness in isolation of each other. The publication of books on happiness has become a lucrative niche market, to be sure.

Pop Happiness

In the twentieth century, Martin Seligman and his colleagues on the beaches of Mexico birthed an entirely new field called "positive psychology," which aims to get us to focus on what is right with our lives rather than what is wrong. Positive psychology studies positive emotions, positive traits, and positive institutions in an attempt to make your life more, well, positive. The idea is to capitalize on your strengths rather than to emphasize your weaknesses or detriments. (To lead a productive and fulfilling life, you can take an online authentic happiness test.)[5] Seligman argues that your happiness is based on who you are intrinsically, voluntary actions, and your circumstances. Tal Ben-Shahar's Positive Psychology class has been and continues to be the most subscribed undergraduate lecture course at Harvard University (maybe because it's an easy A?). Clearly, intelligence and youth don't guarantee happiness.

Sonja Lyubomirsky takes positive psychology even further by breaking the driving forces of happiness into a pie chart: she states that happiness is 50 percent genetics (set point), 40 percent up to your own behaviors, and 10 percent environment (national or cultural region, demographics, gender, ethnicity, experiences, and other life status variables such as marital status, education level, health, and income).[6] More recently studies put the heritability of happiness (i.e., satisfaction with life and well-being) somewhere between 32 and 36 percent.[7] One genome-wide analysis found two genetic variants associated with subjective well-being (i.e., contentment),[8] while yet another report suggests there are at least twenty more,[9] which implies that we won't be genetically engineering happiness very soon. The argument that your state of happiness is only 10 percent based on your circumstances/environment becomes difficult to parse considering that we live in our environments 24/7 and are constantly barraged with commercials of what we need to be happy.

Numerous pop psychology books have popped up, arguably because people want to know how to get happier. Each of these books views happiness as one phenomenon, and most confuse pleasure with happiness. Until you can distinguish the difference between these two emotions, you can't recognize either one as unique and you can't understand, let alone fix, the problem for yourself or for your family.

One Origin of the Confusion

If you google "happiness," here's what you get: "pleasure, joy, exhilaration, bliss, contentedness, delight, enjoyment, satisfaction, contentment, felicity." Note the conflation of the concept of pleasure with the concept of happiness in this definition. Where did this conundrum come from, anyway? Who conflated pleasure with happiness in the first place? And how is it that governments and businesses have been able to harness this confusion for their own purposes? (See Chapters 13 and 14.) Here's one quick and dirty explanation of how words make all the difference. Aristotle argued "the *pursuit of happiness* and the *avoidance of pain* is a first principle; for it is for the sake of this that we do all that we do." Enter eighteenth-century political philosopher–economist Jeremy Bentham. Bentham was a curious fellow hell-bent on quantifying and scientifically explaining individual human experience by constructing a tally sheet of happiness. He might be called the godfather of utilitarianism, the term John Stuart Mill coined in the nineteenth century to describe the philosophy of increasing net world happiness as the primary goal of human existence. Bentham argued that each person should consider others' welfare as seriously as his own. But in the process, Bentham bastardized Aristotle: "Nature has placed mankind under the governance of two sovereign masters, *pain* and *pleasure*, and that just happens to be a fact . . . *benefit, advantage, pleasure, good,* or *happiness,* all of which ultimately comes to the same thing." Under Bentham's rubric, anything that minimized pain and maximized pleasure by its very nature increased happiness. Carrying Bentham's rubric forward into the neuroscientific age, anything that triggers dopamine or opioid release and action (see Chapter 3) would equally qualify as generating happiness.

Even academics have confused the concepts of pleasure and happiness. For instance, the *Stanford Encyclopedia of Philosophy* states that there are two separate "accounts" of happiness: (1) hedonism (maximization of pleasure), and (2) the life satisfaction theory,[10] giving them both equal standing. *What?* Since when is hedonism even in the same room as happiness? Aristotle would be turning over in his grave.

Now that you understand the history of the words themselves, how they have been confused with each other, and how even pop psychologists and Google can't tell the difference, let me now make clear how I am defining them, because the brain science says so. For the rest of this book, pleasure, derived from the French *plaisir* for "to please," is defined as the concept of gratification or reward. The keys to this definition are: (1) it is immediate, (2) it provides some level of excitement or amusement, and (3) it is dependent on circumstance. Conversely, happiness is defined as the Aristotelian concept of *eudemonia*—that is, "contentment" or well-being or human flourishing, or, as in the introductory quote from Yeats, "growth"—physical and/or spiritual. The keys to this definition are: (1) it's about life, not the afterlife, (2) it's not prone to acute changes in one's life, and (3) it is unrelated to circumstance, so anyone can be happy, not just the rich and the powerful.

Unraveling the Threads

These two similar yet conflicting aspects of our neurobiology interact with each other, and it is this interaction that serves as the fulcrum on which our lives, our self-worth, and our internal compasses are balanced (see Chapter 10). Our current collective wisdom does not distinguish between reward and contentment at the etymological level, and fails to acknowledge the personal and societal consequences of mistaking one for the other at the biochemical level. And there are consequences, to be sure. That's what this book is all about. Because chronic excessive reward eventually leads to both addiction and depression; the two most unhappy states of the human condition.

This confusion also belies the basis for many of today's most successful marketing strategies (see Chapter 13). Over the past forty years, the dark underbelly of American enterprise has waged war on the American psyche. City College of New York sociologist Nicholas Freudenberg coined the term "corporate consumption complex" for the six biggest industries that sell us various hedonic substances (tobacco, alcohol, food) and behavioral triggers (guns, cars, energy).[11] Add to that the consumer electronics sector, which further takes advantage of our neurobiology, and wrap it all up in some slick Madison Avenue packaging, and you have an unbeatable recipe for corporate profit. In fact, their recipes are continuing to improve: as the science of reward is elaborated and becomes more precise, new techniques in neuromarketing are now becoming mainstream. And as corporations have profited big from increased consumption of virtually everything with a price tag promising happiness, we have lost big-time. America has devolved from the aspirational, achievement-oriented "city on a hill" we once were, into the addicted and depressed society that we've now become. Because we abdicated happiness for pleasure. Because pleasure got cheap.

2.

Looking for Love in All the Wrong Places

You're probably thinking to yourself, *What makes this guy think he knows what's going on in my mind? I'm in charge of my own thoughts and emotions.* Indeed, you are in charge of your own thoughts, which are yours and yours alone. But you share the process of emotion generation and its experience with every other human on the planet. Your feelings of reward and contentment are just downstream readouts of your neurochemistry.

Before treating obese children, I trained for over sixteen years as a neuroscientist—six years cutting up and studying the brains of rats at the Rockefeller University in New York, and ten more years growing neurons in petri dishes at the University of Wisconsin–Madison and the University of Tennessee, Memphis. These years in the lab afforded me a unique view of the relationship between hormones and behavior. Take a neuron, throw a hormone on it (estradiol, testosterone, cortisol), and watch it go bonkers. Those effects I observed in the dish are the same things happening in your brain right now as you are reading this. You're just a jumble of gap junctions, dendritic spines, axons branching, and synapses forming. Some of these connections happen due to current experience, but many of these brain connections are formed before we are ever born. These processes underlie aggression, passivity, maternal behavior, sexual orientation, and gender identity. Almost assuredly, this is why homosexual and transgender youth can't "behave" their way out of it. They're a result of their own neural connections, the result of what came before. But it derives from the same basic tenet: the biochemistry drives the behavior. Because the biochemistry always comes first.

As a scientist, I don't see behavior or emotion. Rather, I see neural pathways and biochemistry, and it's the point of this book to get you to see them too. You see declining school performance. I see inefficient brain mitochondria. You see the diabetes pandemic. I see liver and muscle fat accumulation causing insulin resistance. You see drugs of abuse. I see presynaptic transporters and postsynaptic receptors. You see teenagers glued to their iPhones. I see dysfunction of their prefrontal cortex, the area charged with maintaining attention. You see economic stagnation and societal unhappiness. I see the limbic system, the primitive part of the brain with neural inputs and outputs that drive everything from joy and elation to depression and helplessness.

You see the result. I see the cause. Treating the result never works; it's too late, the horse is out of the barn. Plus, treating the result just papers over the real problem: the cause is still there. Treating the cause works. But you have to understand the cause before you can treat it. It's like the wasps in your attic. Which is more effective: killing the wasps one by one, or destroying the wasps' nest? You have to work upstream of the problem. Which means we're going to need a very short (I promise) course in neuroscience.

My Brain? That's My Second-Favorite Organ

There are hundreds of brain areas that have evolved to perform different functions. The parietal lobes are where touch is interpreted. The frontal lobes cause muscle movement. The occipital lobes are where we see. The temporal lobes are where we hear. But where do we laugh and cry? Where are joy and sadness and fear and disgust and anger felt? In this book we're going to focus on the limbic, or emotional, brain. This system comprises a set of specialized structures deep within the brain, which are all interconnected. And those connections lead to stereotyped emotions in each and every one of us.

The brain is made up of billions of neurons (nerve cells) that are in constant communication with each other through an elaborate neural network. Each neuron has a cell body that makes proteins so the neuron can stay alive, and neurotransmitters that allow neurons to communicate with each other. They each have dendrites, which are special appendages that receive information, on which there a number of receptors. Neurotransmitters and receptors can be described as floating keys that fit into specific locks. Each neuron also has one long axon, a special fiber that transmits this information. When a neural impulse is generated in the first cell, it pulses down to the end of the axon, which contains little packets of neurotransmitters (the keys) waiting to be released. The firing axon then shoots the neurotransmitters across the synapse to bind to the receptors (the locks) on the dendrites of the next cell.

Throughout this book, we're going to be talking about three specific limbic brain systems (see Figs. 2-1, 2-2, and 2-3).

(1) The first system, the "reward pathway" (Fig. 2-1), is made up of neurons (brain cells that send and receive information) that synthesize

FIGURE 2: The brain's limbic, or emotion regulation, system. The limbic system consists of three major pathways that send and receive chemical information that is translated into positive and negative emotions. The interplay between these three distinct pathways dictates both the perception of emotion and the resultant behavioral responses.

Reward Pathway

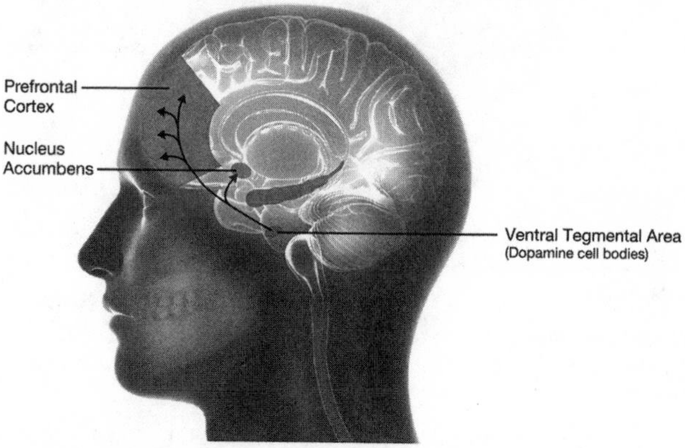

Prefrontal Cortex

Nucleus Accumbens

Ventral Tegmental Area
(Dopamine cell bodies)

Fig. 2-1: The reward pathway utilizes the neurotransmitter dopamine to communicate between the neurons of the ventral tegmental area (VTA) and the dopamine receptors of the nucleus accumbens (NA) to generate the feelings of motivation that attend reward and learning.

the neurotransmitter (chemical for communication) dopamine in a primitive (you don't control it, it controls you) nucleus (collection of like-minded neurons) in the midbrain known as the ventral tegmental area (VTA). When neurons in the VTA fire, they send their dopamine to another brain area called the nucleus accumbens (NA) to generate the feelings of motivation that attend reward. The NA is also a "learning" pathway—learning what feels good (shopping, alcohol, masturbation). Those neurons then release a set of neurochemicals known as endogenous opioid peptides (EOPs), which have the same effects on the brain as morphine and heroin do, and which generate the feeling of pleasure or bliss.

Contentment Pathway

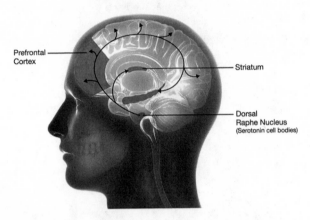

Fig. 2-2: The contentment pathway utilizes the neurotransmitter serotonin to communicate between neurons of the dorsal raphe nucleus (DRN) and multiple sites throughout the cerebral cortex, where the brain interprets impulses as "good" or "bad."

(2) The second system is the "contentment pathway" (Fig. 2-2). A different primitive area in the midbrain called the dorsal raphe nucleus (DRN) contains neurons that produce serotonin and fan out to distant sites all over the cerebral cortex, the thinking part of the brain, where you process experiences and make judgments like "That's good" or "That's bad." Serotonin acts in different ways on different neurons, depending on each neuron's function and the type of receptor (a specialized protein that receives and binds with the molecule, to alter the firing of the next neuron) that sits on its surface.

(3) The third brain system is the "stress-fear-memory pathway" (Fig. 2-3). There are four areas of the brain involved in this pathway. The amygdala is your stress or fear center. It is a walnut-shaped area, one on either side of the brain. When you're in a dark alley, your amygdala is going gangbusters. The amygdala is in communication with three other areas. The hypothalamus, at the base of the brain, controls all the

Stress-Fear-Memory Pathway

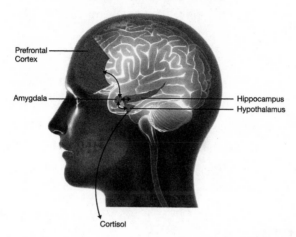

Fig. 2-3: The stress-fear-memory pathway consists of four areas. The amygdala, or your stress center, is in communication with the hypothalamus (at the base of the brain), which controls the stress hormone cortisol. The hippocampus, or your memory center, interprets memories as both good and bad. The amygdala and the hippocampus are reciprocal to each other. The fourth area is the prefrontal cortex (PFC); this is the wise area of the brain that inhibits behaviors that put you at risk. These four areas together keep your outward behavior in check.

hormones of your body, including the stress hormone cortisol, which prepares your body for extreme duress. It also sends messages to your sympathetic nervous system (the fight-or-flight response) to get ready and your vagus nerve (the vegging, chillaxing nerve that slows everything down) to stop firing. The hippocampus is your memory center. It's where you lay down memories, both good and bad. The amygdala and the hippocampus are reciprocal in that when your amygdala figures out that an experience is not a good one, that information ends up stored in the hippocampus ("I've seen this movie before"). The pain of that hot stove from your childhood resides here, as does the last horror movie you saw. And finally the fourth area is the prefrontal cortex (PFC); this is the wise area of the brain that keeps you from doing stupid things again, like insulting your boss, or going to another horror movie. These four brain areas together keep your emotions from overwhelming your ability to think straight and your outward behavior in check.

These three pathways generate virtually all human emotion, and in particular those of reward and contentment. The motivation for reward is experienced when the dopamine signal reaches the NA. A host of different stimuli (power, gambling, shopping, internet, substances) generate signals of reward, but that internal feeling of reward is pretty much the same whatever the trigger. This is why virtually any stimulus that generates reward, when taken to the extreme, can also lead to addiction. You can get addicted to a drug, but you just as easily can get addicted to a behavior, such as gambling or internet use.

Conversely, while experiencing happiness is predicated upon sending the serotonin signal, the actual interpretation of that signal isn't as simple. It also depends on the receptor that is receiving that signal, which changes how you experience it. This is why the positive emotions derived from listening to certain types of music have a different quality from the ones experienced when graduating from college, which are different from the ones triggered by building a home for Habitats for Humanity. And this is very likely why there are so many different definitions of happiness— many different on-ramps, many different roads, many different speed limits—but only one destination for contentment. Other positive emotional phenomena, such as joy, elation, rapture, and the mystical experience, likely take the same roads but end up taking different exits.

The Would-Be Wonder Drug

To convey the nature of, and the difference between, these two emotions of reward and contentment, and how these emotions have been hacked, I must first convince you that that these and other emotions are rooted in the workings of the brain. So let me now give you two examples of the power and scope of these three pathways on your emotional state, and how manipulating them either externally (using a drug) or internally (through a wayward passion) can take a simple positive emotion and turn it into a weapon of mass destruction. I offer you Exhibit A: rimonabant.

If it weren't for reward or the promise of it, we'd all commit suicide due to deep and inconsolable misery. This hypothesis was put to the test a decade or so ago with the release in Europe of the anti-obesity drug rimonabant (Accomplia, Sanofi-Aventis). This medication looked extraordinarily promising right up to and through its approval for general use by the European Medicines Agency (the European equivalent of our FDA). The first endocannabinoid antagonist, rimonabant was the anti-marijuana drug. As it turns out, many of the brain pathways outlined above possess a receptor for tetrahydrocannabinol (THC), the active compound in marijuana. When THC binds to this receptor, known as the CB1 receptor, it heightens mood and alleviates anxiety, which partially explains why people become so giddy when they smoke pot; it also heightens the transduction of pleasure, which explains why so many pot smokers have sex after they partake; and it explains why people get the munchies afterward. Why would Mother Nature put a receptor for marijuana in our brains anyway? As it turns out, we make our own endogenous brain compound, called anandamide, which naturally binds to that CB1 receptor and which keeps us eating and renders most of us semi-functional in social groups, even if we're not smoking pot. But in those who toke, anxiety is thrown to the wind along with every inhibition it suppresses, leaving plenty of room for pleasure.

Rimonabant blocks this CB1 "feel good–munchies receptor." This drug was approved by the European Commission in 2006. As a weight-loss drug, it worked very well and mitigated the co-morbidities of obesity.[1] The data were incontrovertible. Lots of people lost lots of weight. Their appetites went down and they stopped eating junk food. They lost all interest in food; it just didn't provide pleasure anymore. In fact, they derived no pleasure from anything. Rather, their anxiety increased markedly. For five years prior to approval, rimonabant was all anyone in the obesity field could talk about. Rimonabant was going to be "da bomb." And then it bombed: European post-marketing data showed that 21 percent of the people who took it became clinically depressed, and many of them committed suicide. Sure, you will lose weight if you don't get pleasure from eating. But you also lose your motivation for any reward, which means you lose your motivation for life. Suffice it to say, it was rapidly withdrawn from the European market and the U.S. FDA never approved it.

What does the rimonabant lesson teach us? First, it clearly demonstrates that the biochemistry drives the behavior—and the emotions. Rimonabant, by blocking the CB1 receptor, led many into clinical depression, and led to suicide in a few. Anxiety is one of the chief antagonists to both reward and happiness. No wonder marijuana use has skyrocketed, to take the place of alcohol as the preferred method to reduce anxiety,[2] leading to its legalization in many states. After all, which is more dangerous to society at large, marijuana or alcohol? Since most statehouses are now controlled by baby boomers, the switch is not surprising. Second, it shows us that reward-seeking behavior is a double-edged sword. It's the factor that ensures the survival of the species, but it clearly doesn't ensure the survival of any one individual. In fact, just the opposite. Stifle pleasure (with rimonabant) and we plunge into the depths of despair—yet too much pleasure, which we will learn in Part II, is the underpinning of the addiction response, and also can drive us to the depths of despair. You can think of reward as having a bell-shaped curve (see Chapter 3). There's a sweet spot in the middle. Anywhere else on the curve and you play with fire. And third, it shows us that things that interfere with the normal functioning of the limbic system will increase your anxiety, which will secondarily reduce your pleasure; and when pleasure is reduced severely, it can cause depression and even suicide. *No pleasure means no happiness.* Pleasure is the straw that stirs the drink. Happiness is the drink. Anxiety melts the ice cubes. We all need reward, because reward keeps anxiety at bay . . . for a short time. Since the rimonabant debacle, Big Pharma has conducted many forays into blocking the endocannabinoid system, thus far all coming up empty. On the other side, medicinal marijuana dispensaries are popping up like weeds.

Not Feeling the Love?

Now let's turn our attention to a set of emotions with which virtually everyone has some level of experience—a set of emotions that clearly parse the difference between the reward and contentment pathways, and how conflation of the two can get you into some serious hot water. I offer you Exhibit B: love.

"I love you, now let's make love." Two different statements, stemming from different biochemical reactions in the brain, with little in common except the word. Love is the harbinger and result of contentment; sex is driven by our need for reward.

Amoeba engage in asexual reproduction. They don't need a limbic brain system. They don't even need a brain. But mammals do, and they can't do it alone. They not only need a partner, they get off on it. Friction in and of itself is not necessarily a pleasurable experience, but throw in a few sex hormones and you can generate quite a party. And that's why mammals have genital nerves that transmit what would otherwise be an unwelcome sensory annoyance anywhere else on the body as exciting foreplay instead. Even male rats engage in foreplay before copulation and their eventual intromission (the rat version of ejaculation). But this happens only under the influence of the male hormone testosterone. No testosterone, no interest. And female rats arch their backs to attention (known as lordosis) to allow for copulation whenever their flanks are stroked, but only under the influence of the female hormone estrogen. It's the same with humans, minus the lordosis. Think about it. Why in the world would post-pubescent young adults otherwise endure the pain of possible rejection, the idle chitchat, the overpriced bar bill, the smelly pheromones, and the bad breath, if there wasn't a really top-notch reward at the end of it? There had better be a big payout. All because of testosterone and estrogen. Before sex hormones kick in at puberty, it's all cooties. And then it's all angst. Again, the biochemistry always comes first.

Love has been around for all of human history, hasn't it? Maybe not . . . as Tina Turner admonished us back in 1984, "What's love but a secondhand emotion?"—suggesting that it has hardly been a primary endpoint. While this sentiment may not garner me a top spot in the romantic bastion of chick lit, it really is all about the biochemistry.

One of the seminal problems parsing the pathways of love is yet another etymological problem. The Inuit have fifty-six names for snow, but we have only one. Similarly, the Greeks had three words for love, which relate more closely to the biochemistry than our one word. *Eros* is the intense infatuation you feel for your partner at the beginning of the relationship, often tied up in sexuality, and based on testosterone, estrogen, and suspension of reality. It's a hot mix of increased dopamine in the reward pathway, but with a reduction of serotonin in the happiness pathway.[3] Biochemically, it resembles a transient obsession. *Philia* is the more "chill" love you feel toward friends, family members—the love (and frustration) you feel toward your parents and your long-term spouse. The mediator of love between a mother and her children is a different hormone called oxytocin and works through a different emotional pathway unrelated to the three delineated in this chapter.[4] Animal studies where oxytocin is either genetically knocked out or pharmacologically inhibited demonstrate that an otherwise caring mother becomes completely disinterested in her offspring, often to the point of allowing them to starve to death.[5] We will see in Chapter 14 how this pathway can be compromised by dopamine, even in humans. And finally, *agape* is the love one feels for God, but when hijacked by dopamine, resultant zealotry can lay waste to entire peoples in the name of religion (see Chapter 16).[6] So is love an emotion? Or is it just the outward manifestation of some wayward biochemistry? Ultimately there's no difference.

Using new imaging techniques like PET scanning, we can now peer into the brains of people who are in love, i.e., infatuated. In this state, dopamine charges like a bull into the china shop of the PFC, that part of the brain that's supposed to keep you on an even keel, so they become impulsive and aggressive (see Chapter 4).[7] At the same time, serotonin levels fall, reducing any feelings of contentment that might modulate their angst. People suffering from infatuation exhibit anxiety, stress, and obtrusive thinking—the emotions and the behaviors of obsessive-compulsive disorder, a psychiatric condition associated with low serotonin levels.[8]

Love Letters

Now some of you are likely saying, *Wait, it's the relationship, the meaningful connection to a member of the opposite (or same) sex, that makes the process worth it, and which drives the behavior.* Obviously, you've never been on a Tinder date. If you ever thought pleasure and happiness were the same, sex will teach you otherwise. Sex brings reward to a relationship (dopamine) and can mature into contentment (serotonin) if you're lucky. But contentment doesn't drive survival of the species or evolution. Reward does. We don't know if love is uniquely human. We can't tell if rats are happy. We don't even know if primates are happy. We do know they're social. We know they congregate, and we know they have alpha males and subordinates, which is a manifestation of their degree of response to hierarchical stress. We know they demonstrate empathy, which suggests they do have emotions. But we don't really know if they're happy or what makes them happy. Nope, it's reward that drives the primary DNA directive, the survival of the species. Yet the contentment of being in a mutual relationship is a bonus, albeit achievable only for some of us mortals. Consider yourself lucky if you have access to this dividend. Studies of married couples show that the contentment derived from the commitment of an interpersonal union generates added individual benefit: people within such unions tend to live longer and develop fewer diseases than those who have never married (odds risk ratio 2.59) or those who are previously divorced (odds risk 3.10).[9] Because it's actually about the feeling of social connection that comes with marriage. Those with positive social connections and affiliation are happier, and they live longer because of it.

Still, it's hard to dig out and parse the differences between "love" and "in love" from our human history or literature. Did Adam love Eve? I suppose before she ate the apple, you could make a rudimentary case. After all, ignorance is bliss. But afterward? No love lost there. You might try tracing the first literary exhibition of love to Homer's *Iliad*, which hinges on the love between Paris and Helen—although a more careful reading suggests otherwise. After all, Helen was already married, and the only reason Paris was enamored with her was because Aphrodite promised him the most beautiful woman in the realm. Sounds like Paris was suffering from a good case of infatuation rather than love.

Most writers conflate infatuation (or desire) with love on purpose. Infatuation is dopamine, the revving of the reward system, the *motivation*; it's the phenomenon we are drawn to, in part because we can all identify: we've all been there at least once. Infatuation is a big seller, the stock and trade of all those romance novels, and the underlying appeal of *Fifty Shades of Grey* (2011), which has one more kink going for it.

Love, on the other hand, is boring. It's not bad—it's just not a page-turner. Oh, yeah, sure, I know, you're going to throw Erich Segal's *Love*

Story (1970) in my face and tell me that love sells just as well as infatuation. And I'm going to shoot that dead duck down by telling you that virtually every prizewinning novel where love is the basis of the relationship ends up with one of the lovers six feet under. Go ahead, pick your poison. Romeo did. How about Larry McMurtry's *Terms of Endearment* (1975)? Emma gets breast cancer. How about John Green's *The Fault in Our Stars* (2012)? Gus's poison is chemotherapy for his osteogenic sarcoma. Or if one of them doesn't die, then one partner is in some tragic accident and is paralyzed, like in the film *An Affair to Remember* (1957). Nope, you're going to have to work much harder. Infatuation sells because infatuation is reward gone wrong, while long-term love equates with contentment, and requires the protagonist to get killed or maimed to sell books or movie tickets. The stage production of *Into the Woods* (1989) showed us all that "happily ever after" isn't so happy after all.

Perhaps this quote from Louis de Bernières's novel *Captain Corelli's Mandolin* (1994) best explains the difference between "love" versus "in love": "Love is not breathlessness, it is not excitement, it is not the promulgation of promises of eternal passion, it is not the desire to mate every second minute of the day . . . that is just being 'in love' which any fool can do. Love itself is what is left over when being in love has burned away, and this is both an art and a fortunate accident."

Addicted to Love?

Robert Palmer posited in his 1986 video, "Might as well face it, you're addicted to love."[10] Hey, at least 30 million YouTube viewers bought it (those five backup brunettes could addict anyone). OK, if infatuation is mediated by dopamine, then could love actually be addictive? Brian Earp, a neuroethicist at Oxford, also parsed this question by describing these two separate kinds of love. First is what Earp called "acute love," the feeling that characterizes being in love, which is intensely emotional—in other words infatuation, or in the words of romance novels, *desire*[11]—and which leads to alterations in brain chemistry that resembles drug addiction, almost assuredly due to dopamine. And of course, as a kind of addiction, being in love can have severe adverse consequences, just as in true drug addiction.[12] We read in the newspapers of the consequences of rages of passion. The second form of love is what Earp calls a "mature" love and which allows for social growth and cognitive learning,[13] most likely driven by serotonin. Thus love itself, the contentment that truly defines the emotion, is not addicting.

Perhaps marriage is the most obvious institutional example of the achievement of personal happiness today. Yet the role of love in marriage is a relatively new occurrence. Historian Stephanie Coontz states that marriage grew out of convenience to document legitimacy of offspring, obtain the most powerful in-laws to make family alliances, and expand the family labor workforce. After all, even today in some societies, marriages are still arranged by the matchmaker. As Tevye remarks on life in 1905 Russia in the Broadway production of *Fiddler on the Roof* (1964): "Love? That's a new fashion." Marriage today isn't even about marriage but the wedding. When girls envision marriage from a young age, they think about the big day. The ring and the dress. The stress of seating charts and picking the right photographer. The bridesmaids, the DJ, the hot chafing dishes, and the open bar. That's marketing designed for your dopamine system, not serotonin, and not even necessarily directed at the bride. Not enough thought is given to what happens after the credit cards are maxed out and the flowers that match the dresses have wilted. The rock? That's the reward for the effort, not the long-term happiness.

The differences between infatuation and love are but one example of the general dichotomies between reward and contentment. First is the experience of the motivation, of the pursuit; this is often followed by the heightening of visceral reward in the attainment of physical and sexual

responses, often followed by a consummation of the entire reward experience in the form of sexual release. Ultimately over time is built a level of
contentment and bliss that comes from the social and spiritual connection.

Two different neurotransmitters (dopamine versus serotonin), two
different brain areas of residence (the ventral tegmental area versus the
dorsal raphe nuclei), two different targets of action (the nucleus accumbens versus the cerebral cortex), two different sets of receptors, and two
different regulatory systems. But each influences the other, in ways that
can reduce both pleasure and happiness. When taken to their extreme,
these two pathways can take you to the highest mountain or the lowest
valley—addiction, depression, and just plain misery. The science in Parts II
and III says so.

PART II

Reward–
The Agony
of Ecstasy

3.

Desire and Dopamine, Pleasure and Opioids

Contrary to the theme song from the musical *Carnival* (1960), love does not make the world go round. Rather, *Cabaret* (1966) got it right: money makes the world go round. Because money buys prestige and power and sex and big toys. All of these boil down to the same thing: reward. Regardless of the species, the motivation to attain reward (eat, fight, mate) remains virtually intact and unchanged throughout evolution. It makes sense that reward is a pretty strong and unflagging driver of emotion. There are myriad rewards out there, many shaped by culture, religion, and/or reality TV. However, the underlying, unflagging, and omnipresent truth is the drive to attain it. The impetus to get out of bed is generally a quest for reward, whether it's going to work in order to pay the electric bill, or entering the fiery gates of Mordor to destroy the One Ring. For me, it's two cups of coffee.

Virtually every human endeavor is imbued with an inherent reward. Reward has been the predominant driving force since Homo sapiens inhabited the planet. In fact, reward has been a primary driver of personal and collective behaviors since our vertebrate forebears emigrated from the primordial ooze. If we didn't like sex and food, we would never eat anything or reproduce. Reward is how humans (and other species) get things done; it is literally survival of the species.

Manifestations of reward are evident in all measures of personal triumph (that of business titans and/or presidents). Your salary is a general measure of your competence—that is, the reward you provide to others, and that same salary is your reward as well. And manifestations of reward remain the indices of successful companies (quarterly report) and societies (gross domestic product).

Our society does not hurt from the inability to access reward. We've made it our highest priority. Now it's everywhere and ripe for the taking, and virtually nobody needs any extra strategies other the ones they already possess to locate and access it: you need go no further than social media, online porn, your drugstore, your liquor store, or your refrigerator.

Reward is first and foremost. Reward is *the* end. And sometimes reward literally becomes *your* end. Because one reward is never enough. When reward becomes the *primary* goal, overwhelming all else, the end consequence can be addiction—perhaps the nadir of unhappiness. Therefore, understanding the inner workings of reward is paramount to any discussion of personal or societal benefit or detriment.

Can I Get a Double?

The reward pathway is where some of our most basic survival instincts, such as eating and mating, are housed and expressed. The pathway and its mechanisms are thought to have evolved to ensure perpetuation of the species: if there weren't some level of enjoyment to procreation, genes would never get passed on. Despite the varied substances and behaviors that drive reward, the neural pathways and signaling mechanisms are surprisingly similar for all of them. Over the past thirty years, due to the coming of age of some novel biochemical, molecular biological, pharmacological, and imaging techniques, scientists have been able to piece together the drivers and the business end of the reward system, and understand how it can be manipulated for good (and for bad).

Up until recently, the reward pathway was thought to be a one-way express lane to pleasure. But new studies have revealed that the experience of reward is actually two intertwined and conjoined pathways and experiences, with two sets of neurochemicals and two sets of receptors. Although science can piece the two apart, we humans tend to experience them either simultaneously or in quick and rapid succession. The two phenomena can be summed up as: (1) *motivation* or *desire*, mediated by the neurotransmitter dopamine and its receptors. Dopamine is responsible for the outward manifestations of "seeking" behaviors. This is then followed by: (2) *consummation* or *pleasure*, mediated by a class of neuromodulators called endogenous opioid peptides (EOPs, specifically beta-endorphin, enkephalin, and dynorphin) and their receptors, collectively known as opioid receptors. These pleasurable sensations that EOPs generate in the consummation of reward are all experienced inwardly. Thus, on the outside looking in, it's the dopamine effect you see.

While there are several other brain peptides and neurotransmitters involved in facilitating the reward response, for ease of explanation, we can distill the discussion down to the trigger of the pathway: dopamine. Understanding dopamine will be enough to explain how and when we jump the rails. To wit, virtually all pleasurable activities (sex, drugs, alcohol, food, gambling, shopping, the internet) employ the dopamine pathway in the brain to generate the motivation. But too much dopamine starts the downward spiral toward misery. If you can put "-aholic" on the end of the word (alcoholic, shopaholic, sexaholic, chocaholic), then the dopamine pathway is in play.

Dopamine is the fulcrum on which reward tips your scale, or trips your trigger, or floats your boat. The motivation pathway is a conduit between two deep brain structures, the ventral tegmental area (VTA) and the nucleus accumbens (NA) (see Chapter 2). It's a signal from one brain center to another. The cell bodies (the main part of neurons, also known as perikarya) that drive the impulses we experience as motivation are located in the VTA, part of the primitive brain over which you have no control. The VTA serves many purposes, primary among them being dopamine production. These cell bodies then send the dopamine signals to the nerve endings of a second set of neurons that reside in the NA, as well as some others.

When we talk about dopamine and reward, we're talking about the communication between the VTA and NA neurons. The VTA makes the dopamine and sends it across the synapse to the dendrites of the NA. There are other neurotransmitters and hormones involved in modifying the dopamine signal, but we can limit our discussion of motivation to just dopamine without losing anything in translation.

Graded on a Bell-Shaped Curve

Dopamine is a Jekyll-Hyde neurotransmitter. Without it, you're a laconic couch potato; too much and you can get aggressive and paranoid. In other words, like so many things in science and medicine, there is a sweet spot, an optimal level within the dynamic range of experience where the system functions at its best. This can best be illustrated with a bell-shaped curve, which one can travel along backward and forward, depending on your physiologic and emotional state (Fig. 3-1). If you're at the low end (on the left) of the bell-shaped curve, you have little motivation for reward. A slight upswing to the right of a dopamine boost can help you liven up your mood and experience excitement. But if you're already at the top of your

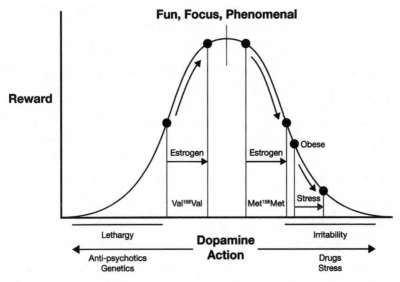

Fig. 3-1: Ring my bell—the curve of reward. The reward pathway functions optimally in the middle of its dose-response curve. Less reward yields lethargy, while more reward yields irritability. Anti-psychotics (e.g., risperidone), by blocking dopamine action, shift the curve to the left, while dopamine transporter blockers (e.g., cocaine) shift the curve to the right. Also, genetic polymorphisms alter your place on the curve. The Val158Val genotype of the dopamine receptor shifts the curve to the left, while the Met158Met genotype shifts the curve to the right. Obese people are right shifted, so more food, meaning more dopamine, confers less reward.

bell-shaped curve, and you get that same dopamine boost, it can result in a new transitional state that can be quite unpleasant. Moreover, your current position on that bell-shaped curve can be changed by your experiences with the many forces, including stresses and medicines, that you are exposed to every day. Let me give you two examples.

(1) **Obesity.** Obesity plays havoc with your dopamine system in very consistent ways. If you're obese, you're already past that central optimum, on the right side of the dopamine curve. Stress will push you even further to the right (see Chapter 4). Then throw in a food cue (an advertisement for

Oreos) and the dopamine in your head becomes so blaring, you have no-where to go but down.[1] The hormone leptin (which comes from your fat cells and tells your brain you've had enough Häagen-Dasz) normally re-duces dopamine firing in the reward center (VTA)[2] and moves you left-ward on the curve (I want ice cream—I ate ice cream—yay, ice cream!).[3] But when your neurons are leptin resistant,[4] as seen in chronic obesity, leptin doesn't work; it can't extinguish that dopamine signal, dopamine action stays high, and you're on to your second, third, and fourth pint—hoping for an ever-dwindling reward.[5] (If you want to learn more about leptin resis-tance and obesity, read my book *Fat Chance: Beating the Odds Against Sugar, Processed Food, Obesity, and Disease*). Furthermore, some people have genetic reasons for their obesity: their NAs are larger, and functional MRI shows that their NAs light up more in response to food commercials than do those people whose weights are normal,[6] thus driving increased interest in food.

(2) Estrogen. At least half of all women will tell you that their menstrual cycles make them hormonal, playing havoc with their level of perfor-mance on simple tasks and their working memory. Rising estrogen means rising dopamine. At the time of ovulation, when estrogen level is at its peak, women can be either focused and motivated, checking things off their to-do lists, or on the verge of maiming their family members for forgetting to pick up the ice cream. Who's who and why? Which are you? Depends on where you start on the dopamine bell-shaped curve, which is likely predetermined by genetics. Around 25 percent of women start on the left side of the curve because they have the Val[158]Val genotype (the combination of genes on each set of chromosomes in each cell) of the protein that chews up dopamine, meaning they have less dopamine hang-ing around, especially in the prefrontal cortex, which is the executive planning and rational part of the brain. When their estrogen rises before ovulation, it actually shifts them to their optimal level on the curve, and they become clearer and sharper.[7] Another 25 percent of women spend the majority of the month at their optimum dopamine levels on the curve. That boost in estrogen at ovulation pushes them farther to the right of the curve, which can cause befuddlement, irritability, and aggression. So if your girlfriend snaps at the smallest provocation on a monthly basis (when she's ovulating, and assuming she's not on birth control), it may be due to her having the Met[158]Met genotype instead.

Get a Hit, Get a Rush

Not only is where you are on the curve important, but how much of a dopamine signal you can generate will also impact your motivation response. There are three separate modes of regulation to the dopamine pathway, and any one of them can go haywire, skewing you to the left or to the right of the bell-shaped curve, affecting your mood and behavior.

(1) **Synthesis.** Dopamine is made or synthesized in neurons of the VTA from the amino acid tyrosine, found in many foods (Fig. 3-2). Ideally, the dopamine concentration in the VTA is tightly regulated and balanced. Too much dopamine can cause a myriad of problems, including psychotic symptoms. Doctors once used drugs to reduce dopamine synthesis in schizophrenic patients. While successful in ameliorating symptoms of outlandish thought, the drugs also caused patients to feel severely depressed, and ultimately these medications were removed from the market. Doctors have also used drugs that increase dopamine production and/or its release in order to treat chronic depression. This has proved to be helpful in some patients, but side effects for others include irritability, aggression, and paranoia. Medications affecting dopamine production are still being researched to help patients find that sweet spot on the bell-shaped curve.

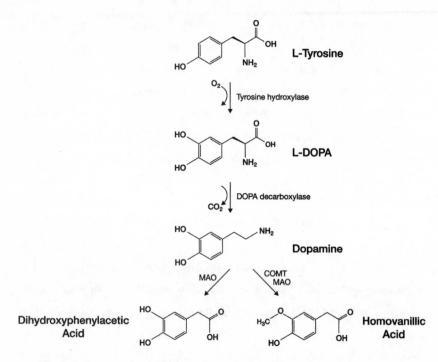

Fig. 3-2: Dopamine synthesis and metabolism. The amino acid tyrosine is acted on by the enzyme tyrosine hydroxylase and receives a hydroxyl group to form L-DOPA. Next, the enzyme DOPA decarboxylase cleaves off a carboxyl group to form dopamine. Dopamine is cleared by the enzymes monamine oxidase (MAO) and catechol-O-methyl transferase (COMT).

(2) **Action.** After dopamine (the key) is released from the VTA axonal nerve terminal, it travels across the synapse, where it binds to a dopamine receptor (the lock) on the NA neuron, and excites it, causing the NA neuron to fire, thus generating reward. The number of receptors determines the magnitude of the reward. More functional dopamines receptors mean more chance that any given dopamine molecule will find a receptor to bind to, and therefore more reward signaling even in the face of less dopamine released. Like extreme couponing, ideally you get more for less. But if the number of receptors is reduced, then each dopamine molecule has less chance of finding a receptor to bind to, and therefore will

generate less reward. This is a non-specific phenomenon known in medicine as the law of mass action,[8] designed to limit each cell's exposure and vulnerability to chronic stimulation (see Chapter 4). It keeps everything in check. Things that change the receptor number, like genetics and drugs, will influence your position on the bell-shaped curve.

Some people harbor an alteration in the gene of their dopamine receptors, making them less able to generate the same level of reward. As an example, Eric Stice at Oregon Research Institute has studied the eating habits of patients who harbor the TaqA1 allelic variation of a dopamine receptor, which means that they possess 30 to 40 percent fewer receptors than the rest of the population.[9] They need more dopamine in the synapse to occupy fewer receptors, so they need a greater amplitude of motivation to derive any reward from it. And as you might expect, their dopamine receptor number inversely relates to their eating behaviors and their weight gain; fewer receptors means more food intake is necessary to generate any reward, and therefore more weight gain. They need more of a fix to generate the same level of reward as people without this particular genetic variation.

Alternatively, the dopamine receptors can be blocked by drugs, so that the dopamine released across the synapse never reaches its target. This is how the dopamine antagonists work. In the 1950s the original anti-psychotic drugs, such as chlorpromazine (Thorazine) and haloperidol (Haldol), revolutionized psychiatry. Up to that point, schizophrenics (1 percent of the population) required long-term or permanent stays in psychiatric wards or care facilities. The dopamine antagonists reintegrated many patients back into society. But these early drugs had severe side effects, such as tardive dyskinesia (uncontrolled movements of the body). The newest generation of antipsychotics, including risperidone (Risperdal), olanzapine (Zyprexa), and aripriprazole (Abilify), have managed to eliminate many of those adverse effects. These medications are often prescribed to adults to enhance the effects of their antidepressants.

They are also prescribed as mood stabilizers in irritable children with aggressive and disruptive behavioral disorders (such as autism, ADHD, obsessive-compulsive disorder, and Tourette's syndrome). But they have some of their own side effects. One possible side effect of their use is a flat affect: they can walk around with little motivation or personality in a Stepford-like haze. These drugs can also induce insulin resistance in the liver, driving insulin levels up, and with it, weight gain.[10] Almost every week in my pediatric obesity clinic, I see a child under ten who started their weight gain only when their doctor placed them on one of these mood stabilizers to prevent classroom disruptions.

(3) Clearance. Dopamine is released into the synapse, where it may or may not occupy its receptor (the key turns the lock; the fewer the receptors, the less likely the occupancy). Party's over, lights out, call the Uber driver; it's now time for the mop-up. The dopamine needs to be cleared out of the synapse, which occurs through one of two mechanisms:

(a) The dopamine molecules can be recycled and used again. They can be brought back to the neuron that released them, repackaged into little storage vesicles, and put back into play for the next party. This is the function of the dopamine transporter, or DAT.[11] Your DATs are akin to the childhood game Hungry Hungry Hippos. They transport and suck dopamine back into the nerve terminal, removing it from the synapse and readying it for the next stimulus. One way to alter the function of the DAT is with various drugs. This is how cocaine acts, by binding irreversibly to the DAT and taking it out of commission. Your first bump of cocaine heightens sensation (kind of like what foreplay does), but it doesn't last very long, leaving you wanting more. The DAT is also where methamphetamine (crystal meth) acts, by fooling the DAT into trying to transport it, instead of the dopamine.[12] Either way the overflow means more dopamine in the synapse, triggering more motivation, more aggressiveness, and more movement. The next time you see someone in the subway snapping their fingers and picking their face, don't ask them how their dopamine is doing—just know that's what's doing it. But the DAT can also be a target of drug therapy for ADD or depression or hypersomnia, as this is where methylphenidate (Ritalin) and bupropion (Wellbutrin) and modafinil (Provigil) also work to increase motivation, but without the face picking.

(b) Alternatively, dopamine molecules can be deactivated by two enzymes called monoamine oxidase (MAO) and catechol-O-methyltransferase (COMT) (Fig. 3-2). These enzymes are your personal Pac-Mans, and they gobble up the dopamine and remove the chemical from the synapse entirely. When the dopamine is either recycled or deactivated, the *wanting* is extinguished. Conversely, dopamine levels in the synapse can be raised by using drugs that inhibit MAO (e.g., phenelzine [Nardil], or one of the original antidepressants), which means less deactivation, more dopamine, more anticipation of reward, and more motivation.

When the DATs and MAO/COMT enzymes (hippos and Pac-Mans) aren't functioning properly due to genetics or illicit drugs, your bell-shaped curve skews to the right. With reduced clearance, more dopamine hangs out in the synapse, meaning more activation at the dopamine receptor and all the baggage that comes with it (see Chapter 5). Unfortunately, your DATs and MAO/COMT enzymes are not very good at determining if you are on your way to, or coming home, from the party. Conversely, if they are too active, they can remove dopamine from the synapse before it ever reaches its destination. Less dopamine, or less binding to receptors, means less motivation and reward.

Too Much of a Good Thing

Recreational drugs, such as cocaine, are the quickest way to boost your dopamine. But drugs aren't the only way to access reward, and drug use isn't the only manifestation of a disordered reward pathway. Humans exhibit a

slew of behaviors that can accomplish the same effect on dopamine transmission, generating the same rush that can be just as acutely satisfying. Unfortunately, some of these can quickly become addictive behaviors, and can
get you into the same long-term kind of trouble. Perhaps the behavior we
have the most data on is gambling.[13] The excitement of the Kentucky Derby
is unmistakable: it's an annual no-miss event at our house. It generates the
same dopamine rush, to different extents, as a ski run down a steep slope, a
shopping spree at Saks Fifth Avenue, or a line of cocaine.[14] One spin of the
roulette wheel doesn't make you a compulsive gambler, just as one bump of
cocaine doesn't make you an addict. But one dopamine rush often turns
into two, and in virtually no time you just might resemble Sky Masterson
from the stage musical *Guys and Dolls* (1950), betting the farm on which
raindrop will reach the windowsill first.

But dopamine is just the gateway neurotransmitter, the trigger. Dopamine is akin to the foreplay before sex (which also releases dopamine): the
experience isn't quite complete until the consummation, the euphoria, the
pleasure—which is mediated through another set of chemicals, the endogenous opioid peptides (EOPs), whose cell bodies are in the hypothalamus, the brain area that controls hormones and emotions.[15] The most
famous of these is beta-endorphin, the brain peptide with properties
similar to morphine. It binds to the same opioid receptor as does morphine or heroin, generating the pleasure signal in the nucleus accumbens
(NA). The opioids are the business end of the reward pathway, and you
can get there with opiate drugs such as hydrocodone (Vicodin) or oxycodone (OxyContin), or with your own beta-endorphin, which is released
in response to vigorous exercise. This is what elite athletes try to achieve
with long-distance running, to get that runner's high.[16] It has been shown
that the pain relief associated with acupuncture is due to EOPs being
released in the reward center. EOPs and opiate drugs bind to their receptors to create the sensation of pleasure. But guess what? Just like with

dopamine, those EOP receptors are also down-regulated with chronic exposure, to limit their action as well (see Chapter 5), although we're not sure what happens with runners.

First the motivation, then the consummation. First the desire, then the pleasure. But that's assuming that your brain already knows what's coming. Our behaviors that typify motivation and consummation are pretty much the same from person to person, but the experiences that trigger them are as individual as you are. What floats your boat may sink someone else's, and vice versa. But you don't know what trips your trigger till you've cocked the pistol. You don't know what you like or want or need until you've experienced it firsthand—at least once. Take a rat naïve to cocaine, morphine, or sugar and implant a recording electrode into the VTA. Then give it a lever for drug administration. Prior to the first ex- posure, the rat doesn't care about the lever and those dopamine neurons are quiet. But after that first hit, the reward signal is registered and that neuron is now primed for action. After that, just provide a cue (like the golden arches of McDonald's) and those dopamine neurons will now fire at fever pitch.[17] The rat will push that lever nonstop.

As a personal example, when I was a pediatric resident, I was hospi- talized for a gastric ulcer after a bout with a toxic chicken curry. The ER doctor gave me a standard dose (15 mg) of meperidine (Demerol), and I was flying. My receptors had never experienced anything like it and had no idea it was coming. I never wanted to come down off that high. Harry Potter must have felt this way when he got his first broom. That single experience explained to me the power these neurochemicals have in shap- ing and changing the motivation and behavior of myriads of people. The problem is that EOPs and their drug counterparts, the opiates, also down-regulate their receptors through the law of mass action. And when opioid receptors down-regulate, you go from *wanting* to *needing*. That's the neurochemical equivalent of addiction.

If You Scratch You'll Keep Itching

The goal of reward is not in the motivation; it's in the consummation. Activating those opioid receptors is where the action is. Pleasure is the goal. Desire is the driver. Motivation drives the outward behavior; consummation is the inward expression of reward.

Let me give you an example of how the reward pathway works *for* us—and *against* us. As I'm writing this, I'm in an Airbnb apartment in Paris (I know, rough draw, but somebody's got to do it), but it's August, it's 95 degrees Fahrenheit with 90 percent humidity, it's a three-hundred-year-old building, and there's no air-conditioning and no ventilation. I'm sticky and I'm stuck in this flat writing, waiting for my wife to return from the Louvre with our kids. Right now I'm thinking of a grande coupe of chocolate ice cream. That's my dopamine telling me to go down to the patisserie on the corner to get a big bowlful, because I deserve it. I could instead get a bottled water to correct my dehydration and to reduce my body temperature somewhat. Cold water can fix my physiology, but I'm not in this for physiology. I'm in this for reward. I'm writing, I'm stressed, I'm hot—and I *really want* some ice cream. I don't *need* it, but I want it—and bad. That's motivation—that's the dopamine talking. I order two scoops—the hazelnut and the pistachio (chocolate is so *Americain*, I decide). Upon my first bite, I get this amazing feeling approaching gustatory nirvana. That's the beta-endorphin, now giving me my food orgasm in my NA. I'm tempted to order a third scoop, but abstain. My wife returns from her excursion through the Renaissance and says, "Two scoops? Really?" At least I didn't get three—would I have gotten 50 percent more pleasure from three than from two? More to the point, did I get double the pleasure from two than I would have had from one? That was the cortisol from the stress, shifting my dose-response curve to the right—a very common sequel to the motivation-consummation paradigm, which yields even more untoward effects than the ice cream itself (see Chapter 4). What is my reaction to my wife's disdain? More cortisol from stress and a rightward shift on the bell-shaped curve. Time for a chocolate croissant.

In summary, reward comes in two phases in tandem: (1) motivation or desire (dopamine from the VTA impacting the NA), and (2) consummation or pleasure (EOPs from the hypothalamus impacting the NA and other areas of the brain). Dopamine is the trigger, EOPs are the bullets. You need both to fire the gun, unless someone else fires the gun for you (like the Demerol in the emergency room). EOPs are also designed to shut down further dopamine transmission to the NA, because, ideally, once your reward has been consummated, you don't need any further anticipation. Cock the trigger, fire the bullet, hit your target, and win the stuffed animal at the fair. Unless . . . you never hit the target. This happens if the signal of either the dopamine or the EOPs isn't effectively transmitted at the NA because of chronic overstimulation and reduction of dopamine receptor number. This leaves you wanting (or even needing) more, and more, and more to get even less of an effect. And the decidedly modern phenomenon impacting our dopamine more than anything else? Chronic stress.

4.

Killing Jiminy:
Stress, Fear, and Cortisol

tress is inevitable. Suffering is not. Your body is built for withstanding acute stresses. Those stresses can be physical (a car accident), adversarial (a lion or a linebacker), medical (high fever), or mental (an English test or forgetting your anniversary). Your body has a protective response to stress, designed to help you fight or flee. It will maintain your blood glucose so that you don't pass out, protect your blood pressure so that you don't go into shock, and prevent inflammation. All of these are mediated by the release of the hormone cortisol from your adrenal glands, which sit on top of your kidneys. You need cortisol, perhaps more than any other hormone, in order to survive; without it, the very thought of getting out of bed is an abomination.

Acute, short-term cortisol release is both necessary for survival and is actually good for you. It increases vigilance, improves memory and immune function, and redirects blood flow to fuel the muscles, heart, and brain.[1] Your body is designed for cortisol to be released in any given stressful situation, but in small doses and in short bursts. Today, even though our acute stresses are declining in frequency and severity (most of us are far less likely to be chased by a lion in our daily lives), our chronic stresses are going through the roof. Despite (or maybe because of) electricity, computers, cars, indoor climate control, and food everywhere, the prevalence and severity of chronic psychological stress and its attendant cortisol effect is taking its toll.[2] I'll lay 20 to 1 odds it's the same for you.

Stressed to the Max

Long-term exposure to large doses of cortisol will kill you . . . but slowly. When pressures are relentless, your cortisol response can remain elevated for days, months, or years. Evidence of the associations between job stress, psychological distress, elevated cortisol, depression, and disease is extremely compelling. Psychological stress in adolescence is directly linked to the risk of heart attack[3] and diabetes[4] in adulthood. Chronic stress also directly impacts the reward pathway as described in Chapter 3, and it has been shown that chronic stress can speed the onset of dementia.[5]

The Whitehall study looked at the health of twenty-nine thousand British civil servants over the course of thirty years. Those lowest on the socioeconomic scale had the highest rates of chronic disease and also of cortisol levels.[6] Even after controlling for behavior (e.g., smoking), death rates were directly related to high stress and the multiple pressures of being financially insecure. Like our friends across the pond, middle- and lower-class Americans also suffer from the highest rates of diabetes, stroke, and heart disease. And if you're not Caucasian, the stresses associated with racism only exacerbate these effects. There are certainly genetic influences, but the fact is that African-Americans and Latinos tend to suffer from higher rates of morbidity from almost every disease than their white counterparts, and stress plays a huge role in this dichotomy.[7,8]

Whatever the mechanism, stress breeds more cortisol. And the more stress, the more breakdown of the endocannabinoid CB1 receptor agonist and anti-anxiety compound anandamide, and the more anxiety.[9] This is where marijuana comes in—to curb the anxiety of everyday living (depending on what you got your marijuana prescription for). Like other drugs, marijuana acts on a specific part of the brain and, depending on whether you are a person who gets paranoid from a few tokes, it can, like, seriously, help you to mellow out. However, chronic marijuana users show long-term cognitive decline to the tune of 8 IQ points,[10] so, in the end, they may be less stressed about reality anyway.

A Bucket of Nerves

The release of cortisol and your body's reaction to stress are the result of a cascade of responses. Threat is first interpreted in a walnut-sized area of the brain called the amygdala (see Chapter 2). Whether it's evading a lion or a line of creditors, your amygdala is scanning the environment for these threats, and talking with other areas of the brain to determine how you should handle it. How your amygdala interacts with the rest of the brain in response to stress determines how you respond, be it curling into a ball or chillaxing. Stress is inevitable. It's your amygdala scanning the scene and how it connects with your other emotions that determines whether you will be safe or sorry. If you don't tame your amygdala early, it can become a devastating creature (see amygdala-taming classes online and Chapter 18).

When a threat is detected by the amygdala, several things occur. First, the amygdala activates the sympathetic nervous system (SNS). The SNS raises blood sugar and blood pressure, to prepare you for the acute stress. Second, like the childhood game of telephone, the amygdala tells the *Hypo*-thalamus (the brain area that controls hormones), which tells the *Pitu*-

itary, which tells the Adrenal glands to release cortisol, known as the HPA axis (like the Gossip Girls). But long-term, this can exact a toll on your arteries and your heart, leading to hypertension and stroke. Third, the amygdala is normally in reciprocal communication with the hippocampus, which is the memory center of the brain. The hippocampus is the set of the Pixar movie *Inside Out* (2015), where memory "bubbles," colored by associated emotions, are stored. The amygdala and hippocampus are supposed to check and balance and exert feedback on each other.

When all is working well, the acute stress you experience is transduced into memories in the hippocampus so that you don't put yourself in the same situation again (a "disgust"-colored memory bubble reminds you that too much tequila leads to an unpleasant morning). Or you are able to realize that, just like last night, the scratching at the door is not a burglar but rather the dog wanting to go out to do his business. Of all the parts of the brain, the hippocampus just might be the most vulnerable to cell death. Almost any brain insult you can imagine (low blood glucose, energy deprivation or starvation, radiation) can knock off the neurons of the hippocampus. And one of the serial killers that attacks the neurons in the hippocampus is cortisol. The longer your cortisol stays elevated, the smaller and more vulnerable your hippocampus gets, which puts you at risk for depression.[11] This is likely why chronic stress is associated with memory loss[12] and why the mothers of toddlers find their car keys in the refrigerator (and not because the kid put them in there).

Executive Function Dysfunction

Excess and chronic stress impacts your ability to reason. The prefrontal cortex (PFC) is your "high order" or "executive function" conscious part of the brain (see Chapter 2). Each of us has our own personal Jiminy Cricket, like the character from *Pinocchio*, which keeps us from indulging

in bad behavior and keeps our baser desires in check. In an uncontrolla-
ble stressful situation, the amygdala–HPA axis commands the release
of neurotransmitters including dopamine (yup, that again).[13] These
flood the PFC, silencing Jiminy, which disinhibits you from doing some
wild and crazy things.[14] When your PFC is under fire by cortisol, your
rational decision-making ability is toast. You can't differentiate between
immediate or delayed gratification.[15] So, instead of your Jiminy telling
you to "Zen" when someone steals your parking space, you are much
more likely to react on impulse and extract your short-lived justice, just as
Kathy Bates's character in the film *Fried Green Tomatoes* (1991) did
(Towanda!).

Worse yet, the more cortisol the amygdala is exposed to, the less it is
dampened down by—you got it—the *law of mass action*. More cortisol
means fewer cortisol receptors in the amygdala, and the more likely your
amygdala will do the talking from here on.[16] Chronic stress day by day
weakens your inner Jiminy[17] to the point where the amygdala becomes
your outer Cricket. You just react to the slightest provocation without any
thought of consequences.

So the amygdala is responsible for your reaction to stress and the release of cortisol. It also interacts with the VTA, the site of the dopamine
neurons. Stress and cortisol also shift your bell-shaped dopamine curve to
the right (see Chapter 3), thereby increasing reward-seeking behaviors.
Increased stress can turn a small desire into a big dopamine drive,[18] which
can be quenched by either drugs or food, or both. This is how the pizza
and beer scenario typifies the American food experience.

Experiments in animals emphasize that either stress or corticosterone
(the rat version of cortisol) administration increases the drive to consume
various drugs of abuse, such as cocaine.[19] One way to drive up the stress
of rats or monkeys is to house them in groups. Invariably one monkey,
through wits, guile, or brute force, will become the alpha male and have
the power to maintain social control over the others, especially in regard

to food and breeding. The alpha's cortisol levels are lower than any other member in the social group. When provided access to cocaine for self-administration, those on the lowest end of the pecking order are the ones that become the addicts.[20, 21] America's middle and lower classes suffer from more chronic stress than the rest of the population: not knowing if there will be sufficient money to pay the rent, working two or more jobs, facing mountains of credit card debt, food insecurity, and a general sense of powerlessness—all ramp up your cortisol. One can argue that this population is at higher risk not only for obesity, heart disease, and stroke[22] but also for drug use and addiction.[23]

Faulty Brakes

Stress-induced dopamine release also has the capacity to remodel the PFC, so now Jiminy isn't even a Cricket anymore; he's been squashed like a bug.[24] These neurons (the ones that house the dopamine receptors) are fewer and farther between.[25] And, if you bombard them enough, you kill them off and they don't grow back (see Chapter 5). You need even more to get less. By driving the stimulation of the amygdala and decreasing your cognitive control centers, stress and cortisol make it much more likely that you will succumb to temptations. Three deep breaths or three doughnuts? Depends on the office you work in.

When cognitive control is lost, the ability to inhibit the drive to seek pleasure is lost. Stress promotes faster addiction to drugs of abuse[26] and is likely the reason why drug addicts find it so impossible to quit. Chronic stress kills off neurons in the PFC, which is a predictor of addicts relapsing.[27] Why do you think rehab centers are generally in scenic areas and designed to be low-stress? It's upon leaving treatment, when addicts are confronted with the stresses of the real world, that some will start using again. This occurs with food as well.[28] Obese people have been shown to have a thinning of their PFCs, likely secondary to long-term dopamine and chronic cortisol bombardment.[29]

And what is America's preferred drug of choice in dealing with stress? The one that is closest at hand. And that would be—you guessed it— sugar. Both animals and humans increase their food intake when stressed or when experiencing negative emotions, regardless of whether or not they are hungry. The boss is yelling at you? Krispy Kreme seems as good a solution as any. And there's actually a reason for this. High-energy dense food, aka comfort food (think chocolate cake)[30] increases acute energy to the brain and thus reduces the amygdala's output and subsequent stress.[31] Stress may affect food intake in several ways. For instance, people with eating disorders tend to show higher levels of cortisol or greater cortisol reactivity.[32]

Alternatively, if stress becomes chronic, and eating is the preferred coping behavior of the individual, then highly palatable food, especially food laced with added sugar, may also become addictive. Cortisol is an appetite stimulant; infusion of cortisol into humans rapidly increases food intake.[33] Those who put out more cortisol in response to chronic stress also consume the most comfort food in response to stress.[34] It gets worse. Cortisol actually kills neurons that help to inhibit food intake. Thus the stress and reward systems are linked, with food (usually sugar) being the drug,[35] breeding a new generation of stress eaters.[36] Break out the Ben & Jerry's.

If Only I *Could* Sleep at the Switch

Another outcome of increased stress is reduced sleep duration and impaired sleep quality, both of which are contributors to, and consequences of, obesity (see Chapter 10). BMI (body mass index) increases over time among short sleepers.[37] A recent study showed that sleep deprivation increases caloric intake by 300 kcal/day, although energy expenditure was unaffected.[38] You're more likely to eat Oreos and watch infomericals about weight loss in those extra hours spent awake than you are to actually work out. Scientists are constantly conducting research to better understand the different parts of the brain. You've likely seen at least one horror or action movie in which a person is strapped to a bed with an electrical helmet, being monitored by men in white lab coats. In real life these types of studies are well monitored for safety and are incredibly valuable in understanding how we tick. When healthy people were asked to spend the night in a lab and deprived of sleep, their brain imaging showed less activation in the PFC but greater activation in the amygdala when choosing foods. Do you think they went for the carrots or the cookies?[39] On nights back in pediatric residency (post-doctoral training where you ostensibly sleep in the hospital), mint Milanos were my only true and unflagging companion that I could always rely on. Poor sleep is common among obese individuals. They often suffer from obstructive sleep apnea (see Chapter 18). They retain more carbon dioxide, which makes them even hungrier, and makes their obesity and diabetes even worse and puts an extra strain on their heart.[40] So stress leads to sleep deprivation. Sleep deprivation causes more reactivity and cortisol release. Cortisol release alters your dopamine response curve. Increased dopamine makes you more likely to eat. The more you eat, the more likely you are to become obese. Obesity leads to sleep deprivation, which can cause stress. A truly vicious cycle.

The Vulnerable Child

Stress and cortisol play an even bigger role in children, a time when eating patterns are programmed.[41] Adverse childhood experiences and early life stressors, such as child abuse, dramatically change the brain,[42] and these changes predispose children to adult obesity and related disorders.[43] Several studies have shown relationships between stress and unhealthy dietary practices, including increased snacking in adolescents.[44] In a study of nine-year-olds, children who felt more stressed by lab challenges ended up eating more comfort food.[45] Not only are stressed kids setups for obesity, they're setups for future abuse of other substances as well.[46]

When you're under stress, your cortisol is up, your PFC is inhibited, your dopamine is firing—all of which will drive you to the chocolate cake or another drug of choice. The more chocolate cake you eat in response to stress, the less pleasure you will get and the sicker you will start to feel, which will drive even more stress. Those dopamine receptors need more, but deliver less. You'll soon become tolerant or, worse yet, addicted.

5.

The Descent into Hades

There's a price to pay for reward. It used to be measured in dollars, pounds, or yen, but now it's measured in neurons. As the monetary price of reward fell, the physiological price of reward skyrocketed. Because those dopamine receptor–harboring nucleus accumbens (NA) neurons are fragile. They want to be tickled; that's why they have dopamine receptors in the first place. But they're very sensitive; they don't want to be bludgeoned. If you open the dopamine floodgates repeatedly, these neurons have some fail-safe methods built in to protect themselves.

The goal is to have an optimal level of dopamine receptors so that even a minuscule dopamine rush (trip to the nail salon) can find an open receptor and generate a unique pleasure from the experience. Obviously, a bigger rush (a couple of glasses of scotch or jumping out of an airplane) will occupy more receptors and generate a bigger reward. But stimulating the dopamine receptors excites that next neuron in the NA, and overstimulation with multiple rapid firings can cause those receptor-containing neurons to go into overdrive, leading to cell damage or death, termed *excitotoxicity*. To protect themselves from your irrational exuberance, each neuron has a built-in subcellular program that reduces the number of receptors on its surface. The more dopamine chronically released from the VTA neuron, the fewer dopamine receptors will be available on the NA neuron to transduce the signal for reward. This is the law of mass action—everything kept in check.

The Firing Squad

When cells in your body die, they are usually replaced with new ones. Not so with the brain, with few exceptions (like the hippocampus, which plays a role in the depression story). Once a primitive brain cell (neuroblast) turns into a functioning neuron and starts firing, it loses the capacity to divide again.[1] Clearly, chronic stimulation of neurons, resulting in cell death with no chance for replacement, is not in your best interest. So nature developed a plan B that is semi-protective. Ligands (molecules that bind to receptors, such as dopamine or cortisol) almost uniformly down-regulate their own receptors, all over the body. In other words, nature makes it so that the locks can be rekeyed or even shut down. But it also means that the response of the cell won't be as robust the next time around. You will need more to get less.

The down-regulation of receptors is a phenomenon called *tolerance*; the receiving neuron is becoming tolerant to the excessive stimulus. This is both good and bad. It's good because it means your neurons aren't dead. It's bad because the next time you go looking, you're going to need more of the substance in order to get the same level of reward. Which ups the ante. Tolerance is a standard response in medicine, and occurs with virtually every chemical that binds to a receptor, whether it be in the brain, the gut, the muscle, the liver, or anywhere else for that matter. Tolerance is how cells, and you, survive.

THE DESCENT INTO HADES

But when neurotransmitters bludgeon the receiving neuron en masse and without cessation, they can overstimulate and eventually kill that neuron through a process of programmed cell death called *apoptosis*. Chronic excitation of almost any neuron can lead to cell death. This is a common phenomenon in neuroscience, and it's a necessary process. Apoptosis is inherent to all cells in the body; it's the self-destruct program that keeps good cells from turning bad (e.g., cancer).

Trimming and Pruning

There are two ways for a cell to die: necrosis (poisoning it from the outside) or apoptosis (self-destruction from the inside). Drugs can do both: they can poison the cell outright, or they can beat it into permanent submission. Apoptosis is a normal and very important process throughout the body that clears away overworked, mutated, or just plain old and decrepit cells. It is especially important during gestation in the womb. We all start out as a single cell—a zygote—and by adulthood we end up as a conglomeration of 10 trillion cells, which have differentiated into hundreds of various cell types along the way. Think of apoptosis as the human equivalent of bonsai, the Japanese art of pruning and sculpting trees to take on new and beautiful dimensions; otherwise they're just gangly weeds. And throughout the various organs of the body—except for the majority of the brain—we maintain the capacity to be able to divide, regrow, and replace cells even as they are cleared away by apoptosis.

Adults appear smarter than toddlers for three reasons: they have more white matter (the fatty part of the brain that insulates neurons and helps transmit impulses faster), which increases information transmission speed; they have a more developed PFC (the executive function center, or Jiminy Cricket, of the brain); and they generally have more experience to draw on. But the number of neurons remains the same. That's why IQ tests can be administered as early as age four. If your four-year-old self could see you now, Marlon Brando's words from *On the Waterfront* (1954) would no doubt reverberate: "I coulda been somebody. Instead of a bum, which is what I am." So keeping your neurons happy and healthy throughout your life should be part of your prime directive.

Yet many of us spend our lives bombarding our synapses with substances and behaviors that down-regulate receptors through the law of mass action (see Chapter 3) or that act as poisons (e.g., alcohol), taking out perfectly good neurons, or with substances that provide different forms of excitation, including illicit drugs of abuse,[2] or too much coffee, too much stress, and not enough sleep.[3] They either rapidly necrose from the outside or, more likely, slowly apoptose from the inside. And then pleasure is strewn to the wind, because those neurons are not coming back. This process is different from developing tolerance. Tolerance is the down-regulation of receptors with the chance of coming back. Once a neuron is dead, it ain't never coming back. There's less to work with, and you can't make more neurons. All of which comes down to the same result: you need more to get less.

As an illustration, let's choose a peanut butter cup, the cheapest of all thrills (but it just as easily could be a shot of espresso or vodka). In terms of the reward neuron, the initial script's the same: Get a desire (dopamine). Get a fix. Get a temporary rush (EOPs). Yum. But, man, that peanut butter cup was so delicious. Just the right amount of peanut butter, salt, chocolate, and sugar. Specifically engineered to hit your bliss point as chronicled by Michael Moss.[4] Go ahead, eat the second one—they come two to a package, after all. Get another rush; this one won't last as long as the first one because there are fewer receptors. Tomorrow, you go get another package at Walgreens—they're staring at you right on the counter—and dig in for your third hit, but you just can't recapitulate that gustatory nirvana again. More should be able to do it: the next day, you buy the six-pack. And now that extra fix means your receptors are down-regulated even more. So you decide to put the pedal to the metal: the economy-size bag has now become your standard, and it's just giving you way less response than you ever had. The cashier at Safeway now recognizes you. And now you have so few reward-transducing receptors, you hardly break a smile. You want the yum, but even after eating the Halloween-size bag and a couple of pints of ice cream while watching *The Notebook* (2004), you still aren't satisfied.

Pickling Your Brain

Every substance and behavior that drives up your reward triggers will just as quickly drive down your reward receptors. Different types of rewards, chronically and in excess, all have the same effect. Why is it that alcoholics can consume so much more booze than your average drinker? Their livers have a much higher tolerance, because repeated and high exposure has increased their capacity to metabolize the alcohol. Their brains also have a higher tolerance because the alcohol has been driving those VTA neurons, which have been bombarding those receptors in the NA with dopamine for so long, they need a lot more to fly the friendly skies. A true lush will get less pleasure from a pint of bourbon than your average normal drinker would get from a single cocktail.

If you stop the nosedive here before your neurons are deep-sixed, you have a fighting chance of pulling out, like a withered flower that's waiting for rain. If the postsynaptic neurons are only damaged but still alive, your dopamine receptors can regenerate over time.[5] You can bring your reward system back and start over, although dopamine receptors aren't back to normal for at least twelve months.[6] One reason for overdose is that addicts, after coming out of a period of abstinence stemming from rehab or jail, will go straight for the last dosage of their drug of choice. Because they no longer have the same level of tolerance, their previous dose is now an overdose.

But keep nosediving and you're sure to crash. Next on the hit parade is the *snowball effect*. Your VTA dopamine neurons, the drivers of the reward signal, are themselves in overdrive. They're working their little nuclei off trying to manufacture more enzymes to make more dopamine (even though your pleasure quotient is almost negligible because those dopamine receptors are so down-regulated). Now both your dopamine neurons and their target receptors are flirting with initiating that apoptotic self-destruct sequence. At some point down the line, they're going to give up. You now have much less of the reward pathway than you used to. You'll never reach the same level of reward as before—ever—because you just don't have the machinery to do so. You're constantly trying to recapture that first high, "chasing the dragon," but you'll never be able to because those neurons are dead. When you fly with one engine, you're much more apt to crash and burn. People in recovery from illicit substances have a motto: "Once a cucumber becomes a pickle, it will never be a cucumber again."

But wait, there's more! A third phenomenon often comes part and parcel with tolerance. Some of our favorite rewards have extra pain built right in. Once you've become tolerant, and you're spending your salary to maintain your fix, you wake up and decide it's time to quit. But changes in those neurons have occurred. The acute cessation of many of these substances can lead to severe and extremely unpleasant experiences, known as *withdrawal*. They tend to cluster as symptoms of physiological withdrawal (e.g., caffeine, alcohol, narcotics, and tranquilizers) with effects on the body, such as sweating, racing heart, palpitations, muscle tension, chest tightness, difficulty breathing, tremor, and GI complaints such as nausea, vomiting, and diarrhea. Some, like delirium tremens (DTs) from alcohol withdrawal, or hallucinations from benzodiazepine (benzos) or barbiturate (downers) withdrawal, can be life-threatening. Or symptoms can cluster as manifestations of emotional withdrawal (e.g., from cocaine, marijuana, and ecstasy), with effects on the brain such as anxiety, restlessness, irritability, insomnia, headaches, poor concentration, depression, and social isolation.[7] These symptoms of withdrawal can be so severe as to prevent people from even wanting to give up their drug/behavior of choice, and they often lead abusers to relapse. Tolerance and withdrawal are the classic two-headed hydra of the definition of addiction.

An Equal Opportunity Offender

When there is too much dopamine, there can be too much motivation to obtain your pleasure. The motivation—the *wanting*—becomes more of a *needing*. Many addicts commit a host of crimes to obtain their fix, often hurting loved ones in the process. They drive drunk, lose custody of their children, and not infrequently become destitute. It can happen to anyone. A friend of mine recently had back surgery and was prescribed oxycodone (OxyContin). Within a brief time period, she stopped needing the drug to ameliorate her physical pain, but she still needed it to occupy those few remaining opioid receptors. She became a drug-seeking doctor shopper, doing whatever she could to fill her next prescription. The pain she suffered was not from her back but rather from the desperation to obtain her next fix. Her dopamine put her in 24/7 motivation overdrive, while her oxycodone didn't pack the punch she needed to calm herself down. We often hear of famous people entering the Betty Ford Center to dry out. Did they get addicted and then get famous? No, drug abuse is a response to the stresses of everyday life. And apparently famous people have stresses too.

People often say addiction is a choice—after all, Nancy Reagan argued that you could "just say no." And despite overwhelming evidence that nicotine caused both tolerance and withdrawal, the tobacco industry used "free choice" as its cornerstone defense from the 1960s through the 1990s. In 1994, on national television, Thomas Sandefur (CEO of Brown & Williamson), William Campbell (CEO of Philip Morris), and James Johnston (CEO of R.J. Reynolds) all testified under oath "I believe that nicotine is not addictive."[8] So how did the tobacco companies get away with saying nicotine was not addictive, and for so long? What were the criteria? Well, they couldn't deny tolerance and withdrawal, so they played down the concern by equating it with other hedonic substances. According to the tobacco companies, "Addiction is an emotive subject and it is certainly possible to define the term broadly enough to include smoking . . . the current definition is more colloquial . . . and certainly applies to the use of many common substances that have familiar pharmacological effects to cigarettes, such as coffee, tea, chocolate and cola drinks."[9] Hey, don't pick on us, we're no worse than Coca-Cola!

Whether they knew it or not, they actually got it right. Individuals in society have found pleasure in all forms of reward—to each his own. Some of those are substances. And some are behaviors, which we engage in specifically because they feel good. It doesn't matter: the final common pathway is dopamine. Some behaviors are innate, like eating and sex. Some are learned, like shopping or shoplifting or gambling or gaming or texting or bingeing on Netflix (see Chapter 14). Similar to taking drugs of abuse, overperforming each of these behaviors can also manifest the phenomenon of tolerance (i.e., performing the behavior more and more to get less and less reward). Some will be so driven by their dopamine to greater and greater extremes to get that ever-diminishing EOP rush that they will escalate up to deviant behavior, some of which are severe enough to put you face-to-face either with the law or with your maker.

How are these behaviors, as bizarre as they are, considered addictions? While they clearly manifest tolerance (more and more for less and less), they do not demonstrate either the physiological or emotional consequences of withdrawal upon cessation.[10] Tolerance without withdrawal—is that addiction? The American Psychiatric Association (APA), the professional society of psychiatrists, are the gurus in this field. They are the publishers of the definitive *Diagnostic and Statistical Manual* (*DSM*), which defines and categorizes all psychiatric and behavioral disorders.[11]

Are Addictive Behaviors Really Addictive?

The *DSM* has undergone various revisions in recent years to keep up with the field, particularly around the concept of addiction, the criteria for its diagnosis, and what it takes to be addicted. There have been two main obstacles in advancing the field of addiction, and both of them have to do with definition and criteria. The first is: Can you be addicted to something that isn't a substance? The second is: If you demonstrate tolerance but not the physiological symptoms of withdrawal, can you still call it addiction? Answering these two questions has taken almost two decades of research and debate—the time between the publication of the *DSM-IV* (1994) and the *DSM-V* (2013). And I promise you, by the time the *DSM-VI* rolls around, there will be further modifications. This field is in constant flux.

For decades the APA said no, behaviors like gambling weren't manifestations of addiction because the definition was tolerance *plus* withdrawal. Lack of withdrawal meant they didn't meet the criteria. But after decades of discussion, policy making, and politicking, the *DSM-V* has removed the requirement for withdrawal as an absolute diagnostic criterion. In so doing, the APA has now changed the definition of substance-related and addictive disorders and allowed for the inclusion of addictive behaviors as well. Here is the current mix-and-match list of eleven items:

1. Tolerance
2. Withdrawal
3. Craving or a strong desire to use
4. Recurrent use resulting in a failure to fulfill major role obligations (work, school, home)
5. Recurrent use in physically hazardous situations (e.g., driving)
6. Use despite social or interpersonal problems caused or exacerbated by use
7. Taking the substance or engaging in the behavior in larger amounts or over a longer period than intended
8. Attempts to quit or cut down
9. Time spent seeking or recovering from use
10. Interference with life activities
11. Use despite negative consequences

Instead of hard-and-fast criteria, this *DSM-V* paradigm allows for scaling of severity. Two or three of the above symptoms indicate a mild disorder, four or five symptoms indicate a moderate disorder, and six or more symptoms indicate a severe disorder.

One question that people always ask: Is there an addictive personality? What they really want to know is if addiction is genetic. There are a lot of children of alcoholic parents who are worried that they will suffer the same fate. Many people are exposed to alcohol and they don't get addicted. Many people (like me) have received the narcotic meperidine (Demerol) as pre-op for a surgery, and they don't turn into heroin addicts. They want to know: If I'm not addicted now, I'm out of the woods, right? Is addiction driven by genes or by the substances themselves? There is no doubt that there are certain genes that predispose people to alcoholism[12] or smoking,[13] but they all impact dopamine in some fashion. If a genetic defect or alteration reduces the number of dopamine receptors, motivation for reward will be increased (see Chapter 3). But there's no gene identified to date that is 100 percent predictive. If you harbor a genetic variation of your dopamine receptor, you do have an increased relative risk,[14] but it's not a faît accompli.

Another issue that has plagued research on addiction is the question of cause and effect. Clearly dopamine neurotransmission is associated with tolerance and withdrawal, but which comes first? Is it that dopamine drives the addictive behavior, or is it the addictive behavior that results in the changes in dopamine? A recent study looked at patients with Parkinson's disease, which occurs due to the degeneration of dopamine neurons in another brain area, the substantia nigra (SN), which controls movement. Parkinson's disease patients experience severe rigidity and tremor, interfering with every aspect of their lives. The neurons in the SN that produce dopamine are not just dysfunctional; they're dying. Parkinson's patients are given drugs, such as L-DOPA/carbidopa (Sinemet) and bromocriptine (Parlodel), that increase or mimic natural dopamine signaling to restore movement. Many people have heard of L-DOPA because of the movie *Awakenings* (1990), based on the work of the late neurologist Dr. Oliver Sacks. These are not drugs that are abused for their pleasurable properties. But these drugs are not specific for the areas that affect movement; they also interact with the dopamine receptors in regions that affect reward-related signaling. It turns out that these drugs drive a panoply of behaviors as unwanted side effects, including aggression, paranoia, and poor impulse control.[15] Some patients have even become compulsive gamblers. Activation of the dopamine receptor means the motivation for reward is enacted, with all the positive and negative consequences that come with it. What these studies show is that the dopamine comes first: the drugs drive the dopamine signal, and the dopamine signal eventually drives these behaviors.

Addiction Transfer

What happens when, for one reason or another, you can't access your favorite fix? Once your dopamine pump is primed, it's just waiting to be fired, for something—anything. People abstaining from one substance will frequently find themselves embroiled with another drug or activity (sex, gambling) that can generate the same effect. No AA meeting is complete without coffee or Rockstar, cookies, and smoking out back. Because once you're addicted to one substance and your dopamine receptors are down-regulated, you can easily become addicted to other substances as well. This is known as *addiction transfer*.

Addiction transfer is a standard alternative: when the addiction you have becomes unacceptable to yourself, your spouse, or society, you move on to the next. A rational person would opt to switch from more addictive, dangerous, and societally eschewed substances to socially acceptable alternatives. It's common for those quitting smoking to start overeating, and most will inevitably gain some weight. Many people have experienced the phenomenon of addiction transfer—for example, when people switch from cigarettes to food ("I have an oral fixation"). William F. B. O'Reilly, a Republican advisor on Long Island (not *that* Bill O'Reilly), experienced addiction transfer firsthand and wrote about his experience in *Newsday*: "Off Sugar, and Wanting to Tear My Eyes Out."[16] O'Reilly first started out hooked on cigarettes, then he switched to alcohol, then he switched to sugar. But then his waistline grew, and finally there wasn't anything left to switch to. In fact, one of the early treatments for obesity was to take up smoking. When you're addicted to one substance and you find yourself abstaining, your dopamine's modus operandi is to find a substitute trigger.

Bariatric surgery, including lap band surgery, reduces the amount of food one is able to consume at any one time. You simply don't have the space in your stomach; you can't eat like you used to. But many of those undergoing the procedure had unhealthy addictions to food in the first place; sacks of peanut butter cups generated the same type of fix for them as heroin might for someone else. Their dopamine pumps were primed and ready. So what do they switch to? Alcohol, the liquid drug.[17] Carnie Wilson, a self-described food addict and former singer in the band Wilson Phillips, underwent gastric bypass surgery in 2000, losing 150 pounds and landing a gig as a pinup model for *Playboy*. She's become somewhat of a poster child for the concept of addiction transfer, as she then found refuge in alcohol as opposed to food.

The Real Thing

A perfect example of addiction transfer, with long-lasting effects for the entire world, was John Pemberton. He was an Atlanta pharmacist and in 1886 he invented the formula for a very special and quite unique carbonated beverage. On May 29, just three weeks later, Pemberton placed the first advertisement in the *Atlanta Journal* for his soft drink (which wasn't so soft in those days), which would from that day forward be known as Coca-Cola. The story of Pemberton and Coca-Cola is widely known, the stuff of urban legend. Back then, carbonation was a big deal, requiring special high-pressure jets to force enough carbon dioxide into a solution. There was no method for reinforcing standard glass bottles, so carbonation had to be done in pharmacies with special equipment and drunk on-site. This became known as the *soda fountain*. Thus, Coca-Cola was originally sold only in pharmacies. But there was another reason as well.

What is not widely known is that Pemberton was a morphine addict, after being wounded in the Civil War.[18] The reason he developed his sacred formula was a long-standing attempt to wean himself off his addiction. But his addiction was ruining his profits, his business, and his life. He spent the next twenty-one years trying to come up with an opium-free painkiller. He went through several iterations, without success. Ultimately, he developed a concoction that included cocaine, alcohol, caffeine, and sugar. Four separate hedonic substances, four somewhat weaker dopamine/reward drugs, to take the place of one very strong one.

Pemberton mixed the four with carbonated water (thought to have its own hedonic properties). However, due to the temperance movement that overtook the South in the late 1800s and due to many Civil War veterans developing alcoholism, he removed the alcohol, and voilà! However, in 1888, Pemberton sold the formula and the rights to Atlanta businessman Asa Candler for a mere $2,500, and Candler proceeded to turn Coca-Cola into the most famous brand in the world. Why so cheap? you ask. Because Pemberton needed the money—bad—and you can guess why. He was sick, addicted, and penniless. He never did beat his morphine addiction, and he died the same year, at age fifty-seven, in severe pain. Not surprisingly, if you go to the Coca-Cola Museum in Atlanta, this sordid story is nowhere to be found.

In 1903, the federal government required the removal of cocaine for public sale, leaving only caffeine and sugar. Were these two substances alone enough to maintain the hook? Of course: Why do you think Starbucks sells Frappucinos? Candler saw his Coca-Cola placed into pharmacy soda fountains all over the country. It is now available in 208 out of the 209 countries in the world (only North Korea is Cokeless; Myanmar capitulated in 2012, and Cuba in 2015) and is by far the world's most recognized brand. And for good reason: it's a delivery vehicle that mainlines two addictive compounds straight to your nucleus accumbens.

Sugar just happens to be the cheapest of our many substances of abuse. But all of these substances do essentially the same thing. By driving dopamine release, they all acutely drive reward, and in the process they also drive consumption. Yet, when taken to extreme, every stimulator of reward can instead result in addiction. For heroin or cocaine, you need a dealer and a wad of cash. For alcohol or nicotine, you need an ID. But for sugar, all you need is a quarter or a grandma. Sugar is the cheap thrill, the reward everyone on the planet is exposed to, the reward everyone can afford. Everyone's an addict, and all your relatives are pushers. And it's only one of two addictive substances that are legal and generally available (the other one being caffeine). That's why soda is such a big seller: it's two addictive substances rolled into one. Everyone has become a willing consumer of the two lowest common denominators. Sugar and caffeine are diet staples for much of the world today. Coffee is the second most important commodity (behind petroleum), and sugar is fourth.[19] Sugar being a primary example, substances have been purified and mainlined, straight to your dopamine receptors. If you don't exercise caution, you'll blow your neurons out.

6.

The Purification
of Addiction

Substances of abuse used to be scarce—a luxury for most of us—and dopamine was at low ebb. Prior to the eighteenth century, virtually every stimulus that generated reward was hard to come by, due to either its scarcity or its expense. You had to go out of your way to obtain the various illicit drugs. There were no stores, no internet, and there was very little porn. We've always had gambling and prostitution, but they weren't on every street corner. Hedonic substances were once rare, limited to alcohol from the Triangle Trade, which allowed for the transfer of slaves from Africa, sugar and rum from the Caribbean, and money from New England.[1] Slowly but surely, advances in technology, commodity crop farming, and globalization have made various rewarding substances readily available, and the ability to engage in rewarding behaviors not just possible but almost constant. Pleasure is now easy and cheap, if nothing else. In the twenty-first century, substances of abuse have become easier and cheaper to obtain all over the world. Whereas these substances were once something to savor and ponder, now they come a dozen to a box (either doughnuts, or beer, or both if you're Homer Simpson).

A Brief History of Addiction

When did substances of abuse first appear on the scene?[2] Archeological digs support the contention that Central Asia's Yamnaya people (one of the three tribes that founded European civilization) had discovered and were trading cannabis as early as ten thousand years ago.[3] The first literary reference to recreational drug use was from 5000 B.C.E., when the Sumerians chronicled the use of opium.[4] The first reference to alcohol in the form of wine goes back to 4000 B.C.E., and the first mention of commercial production dates to 3500 B.C.E. in the form of an Egyptian brewery.[5] But addiction didn't really become a societal problem until we started purifying these substances. The first reference to addiction comes from China at around C.E. 1000, when opium became widely used.[6]

In Western society, however, addiction and addicts remained a relative rarity through most of the second millennium. We've had wine since the time of the Romans, but we had to rely on natural fermentation for its production. In early times wine spoiled rapidly, because early vintners couldn't get the alcohol content up past 5 percent, just like beer, which was equally likely to spoil. Although commercial beer production dates back to European monasteries in the seventh century C.E., succumbing to alcohol addiction wasn't an option: the alcohol content was just not high enough. Alcoholism remained a matter of availability. Once it could be easily bottled, we were awash in hard spirits. Distilled alcohol became the obvious choice of most addicts, because you could ferment and distill just about anything.

Alcoholism became a major societal problem throughout Europe in the 1700s once it became available and cheap. Prohibition turned out to be the anvil on which our current American society was forged. If anything, the dopamine rush from alcohol was increased tenfold by the fact that it had to be consumed in backroom speakeasys. It's not an accident that 1933 saw the passage of the Twenty-First Amendment, which just happened to coincide with both the nadir of the Depression, and with Franklin Roosevelt's New Deal of 1933. The government needed the tax money. But despite our affinity for alcohol, the dopamine rush still remained a luxury, out of the reach of most people, either due to religion, morality, reputation, or expense.

A Cheap Shot

Times have changed. Currently, the National Institute for Drug Abuse (NIDA) puts the U.S. binge-drinking rate at 30 percent for men and 16 percent for women, while alcoholism rates are 9.5 percent for men and 3.3 percent for women.[7] Considering the use rate for alcohol is 67 percent of the U.S. adult population, that means that between one-quarter and one-half of Americans are binge users. That's a pretty high take rate. And this is worth $212 billion in annual revenue to the U.S. alcohol industry.[8] Kids today aren't just bingeing on alcohol, they're also popping uppers, downers, and everything in between. In adolescents over the last thirty-five years, the binge-drinking rates, as well as use of virtually every other illicit substance, has continued to increase.[9]

Alcohol is but one example of substances being purified and manufactured to suit the whims of societal addiction and bludgeon our dopamine receptors into submission. Marijuana is bred to be stronger than ever before, the coca leaf continues to provide both line cocaine and its cheaper cousin, crack, and opium poppies are still grown to make heroin. There's big money to be made. Just ask Walter White from *Breaking Bad* (2008–2013). Those who distill, bottle, and sell these substances know what they're doing and how to capitalize on our dopamine pathways (see Chapter 3). The pharmaceutical industry has made some incredible strides in recent decades, and medications now exist for a series of diseases and disorders that previously went untreated. However, these medicines are also used off-label, masquerading as "cognitive enhancement."[10]

The Other White Powder

It's no secret there is big money in the pharmaceutical industry. The annual profit margin of Big Pharma is 18 percent, with five companies making 20 percent or more.[11] But even this profit margin is minuscule when compared to the money being made on the cheapest thrill possible. The processed food industry grosses $1.46 trillion, of which $657 billion is gross profit, for a gross profit margin of 45 percent. And what drives such profits? The drug that isn't a drug. Or is it? In America, circumcision of males at birth is relatively common. When the Jewish mohel (trained circumciser) performs the ritual called the Brith Milah, what alleviates the pain? He dips the pacifier in wine. But when the obstetrician performs this procedure in the hospital, what alleviates the pain? The pacifier is dipped in Sweet-Ease (a 24 percent super-concentrated sugar solution)[12] that activates both dopamine and opioids in the brain.

Just as we all have motivation to obtain pleasure, virtually all humans have a sweet tooth at some level. It's inscribed into our DNA. The world loves sugar. There's not a race, ethnic group, or tribe on the planet that doesn't understand the meaning of "sweet." This can be traced back evolutionarily, because there are no foodstuffs on the planet that are both sweet and acutely poisonous. Sweet meant that it was safe to eat. Jamaican ackee fruit, when immature (and not sweet), contains a compound called hypoglycin that can cause Jamaican vomiting sickness and can be life-threatening. But once the mature ackee fruit blooms, all the hypoglycin is metabolized, and it is the Jamaican national dish, canned and shipped worldwide.

Despite our sugar love, the cost of sugar prevented its overconsumption until about fifty years ago. Prior to World War II, sugar was a condiment, something you added to your coffee or tea—"one lump or two?" But shortly after World War II, refined sugar became the drug of abuse for the masses. It was ratcheted up first with the advent of processed foods, which included added sugar. Then it was given another hike with the advent of high-fructose corn syrup in 1975, which provided competition for cane and beet sugar. This lowered prices further, and suddenly sugar started appearing in everything. And finally, the first Dietary Goals for the United States,[13] published in 1977, told people to eat less fat, but it didn't say anything about sugar. Now we have a choice: we can get our fix either from cane or beet sugar, or its cousins high-fructose corn syrup, maple syrup, agave, and honey. There's a quick fix waiting for you on every street corner and in every refrigerator.

Rats, like humans, love sugar. Feed them a little, and they will want more. Let them at it, and they will increase their intake, drinking loads of sugar water just to maintain their fix. Columbia neuroscientist Nicole Avena showed that within just twenty-one days, their NA looks whipped, as would happen with any other drug of abuse.[14] And even more so when given binge access.[15]

Until recently, scientists were locked in a vehement debate as to whether it was sugar or fat that caused your reward pathway to fire. Eric Stice in Oregon conducted a neuroimaging study that looked at fat and sugar separately, and together, in milk shakes that were calorically equivalent.[16] Using four different combinations of high- and low-fat and high- and low-sugar milk shakes, he found that the high-fat variety caused greater activation in oral somatosensory regions (i.e., where you experience mouthfeel), while high-sugar more effectively recruited reward-related and gustatory (taste) regions. Increasing sugar caused greater activity in the reward pathway, while increasing fat did not. In other words, it's the sugar that drives the dopamine, which drives the motivation for reward.

Dietary sugar is composed of two molecules: glucose and fructose. Glucose is the energy of life. Glucose is so important that if you don't consume it, your liver makes it (gluconeogenesis). Conversely, fructose, while an energy source, is otherwise vestigial; there is no biochemical reaction that requires it. Yet, when consumed chronically and at high dose, fructose is similarly toxic and abused.[17] Not everyone who is exposed gets addicted, but enough do to warrant a similar discussion.

These two molecules, glucose and fructose, activate different parts of the brain. On functional MRI (fMRI) scanning, the glucose molecule lights up areas associated with consciousness and movement, while the fructose molecule lights up the reward pathway and several sites in the stress-fear-memory pathway as well.[18] These studies suggest that sugar is uniquely capable of driving the reward pathway and altering emotional responses.

Denying the Obvious

Not everyone subscribes to this expanded view of addictive substances. Drugs are a luxury. Food is a necessity. How can a food—like sugar—that is necessary for survival also be addicting? Because certain "foods" are not necessary for survival. We *need* essential nutrients that our body can't make out of other nutrients, or we get sick and die. But there are only four classes of essential nutrients: (1) essential amino acids (nine out of the possible twenty found in dietary protein), (2) essential fatty acids (such as omega-3 and linolenic acid), (3) vitamins, and (4) other micro-nutrients, such as minerals. Just add water and stir. None of the foods that contain these essential nutrients are even remotely addictive. Of those substances that also contain calories, only alcohol and sugar have been shown to be addictive. The other addictive consumable found in food is caffeine.

Wait, I can hear it—you're saying, sugar? A drug? How is that possible? It's part of other foods. It's in fruit. It provides calories. It's an energy source. Moreover, it's a commodity! We subsidize it! OK, let's try an analogy. Can you name a substance that: (1) has calories, (2) is an energy source, (3) is not required by any biochemical reaction in the body, and (4) is not nutrition by anybody's estimation, (5) when consumed in excess causes damage to cells, organs, and humans, (6) we love anyway, and (7) is addictive? Answer: *alcohol*. Alcohol's got calories and is an energy source, but alcohol is not a food. Alcohol is not nutrition. There's no biochemical reaction that requires it (40 percent of Americans don't consume alcohol, and they're not sick).[19] Alcohol does not hurt you because it has calories or because it can cause weight gain. Alcohol is dangerous because it's alcohol. It can fry your brain and your liver. It's a drug and it's addictive in a percentage of the population.

Same with sugar: it meets each one of these same criteria. Fructose, the sweet molecule in sugar, contains calories that you can burn for energy, but it's not nutrition because there's no biochemical reaction in any eukaryotic (animal) cell on the planet that requires it. It's a vestige from when we split off the evolutionary tree from plants. And when consumed in excess, sugar fries your liver, just like alcohol.[20] And this makes sense, because where do you get alcohol from? Fermentation of sugar: it's called wine. Sugar causes diabetes, heart disease, fatty liver disease, and tooth decay. Sugar's not dangerous because of its calories, or because it makes you fat. Sugar is dangerous because it's sugar.[21] It's not nutrition. When consumed in excess, it's a toxin. And it's addictive. Fructose directly increases consumption independent of energy need.[22] Sucrose establishes hardwired pathways for craving in these areas that can be identified by fMRI.[23] Indeed, sweetness surpasses cocaine as a reward in rats.[24] Animal

models of intermittent sugar administration induces behavioral alterations consistent with dependence, i.e., bingeing, withdrawal, craving, and
cross-sensitization to other drugs of abuse.[25]

The naysayers will still say, "But sugar is natural. Sugar's been with us
for thousands of years. Sugar is FOOD! How can food be toxic? How
can food be addictive? This begs the question: What *is* food? *Is* sugar
food? *Webster's Dictionary* defines "food" as "material consisting essentially of protein, carbohydrate, and fat used in the body of an organism to
sustain growth, repair, and vital processes and to *furnish energy*." Well,
sugar furnishes energy, so of course that makes it a food! For instance, a
group of European academic researchers have joined forces into a political action group called NeuroFAST, which maintains that humans can
succumb to "eating addiction" (it's the person's fault) but argue vehe-
mently against the concept of "food addiction" (it's the food's fault).[26]
This is not a semantic difference. If it's "eating addiction," the food industry bears no responsibility. If it's "food addiction," they bear at least
some corporate responsibility for our current medical and behavioral
health debacle. NeuroFAST categorically insists that even though specific foods can generate a reward signal, they can't be addicting, as they
are necessary for survival. In their own words:

> In humans, there is no evidence that a specific food, food ingredient
> or food additive causes a substance based type of addiction (the only
> currently known exception is *caffeine* which via specific mechanisms
> can potentially be addictive) . . . Within this context we specifically
> point out that we do not consider *alcoholic beverages* as food, despite
> the fact that one gram of ethanol has an energy density of 7 kcal.[27]

Interesting. NeuroFAST recognizes caffeine as addictive, yet they give
it a pass. Caffeine is present in many foods (e.g., coffee), yet it is classified
by the FDA as a food additive. It is also a drug; we give it to premature

newborns with underdeveloped nervous systems to prevent apnea (stoppage of breathing). NeuroFAST then goes on to give alcohol a pass as well. Natural yeasts constantly ferment fruit while still on the vine or tree, causing it to ripen,[28] yet NeuroFAST recognizes that purified alcohol is not a food. Alcohol is also a drug; we used to give it to women to stop premature labor. It is also addictive.

So what is the difference with refined sugar? The sucrose in your sugar bowl is the same compound as what is in fruit, but the fiber has been removed, and it's been crystallized for purity. It's this process that turned fructose from food into drug. And it's the purification that made it addictive. Just like alcohol. So in a convoluted sort of way, NeuroFAST got it half-right. They state that food can't be addictive, but they recognize that food additives can. But that means when these additives are added to our food, they can make our food addictive. Like sugar.

The sine qua non of this argument is soda. Is soda a food? Is there anything in soda that you need that could make it a food? Sugar—that's an additive. Caffeine—another additive. Both addictive. Phosphoric acid, caramel coloring? No. Sodium? We're all consuming triple what we need as it is.[29] Water? Water's necessary, but it's not a food—it's water.

In the Middle Ages, sugar was a spice. Up through the mid-1900s it was a condiment. Only in the last fifty years has it become a diet staple. And it's addictive for exactly the same reasons and via the same mechanism as alcohol. Sugar is not a food; it's a food additive, just like alcohol. That's why the FDA proposed changing the Nutrition Facts label to include "*added* sugar" (although the current administration may revoke this change). And that's why children are getting the diseases of alcohol—type 2 diabetes and fatty liver disease—without alcohol. And that's just what it does to your body and your reward pathway. Hold on, the party's just getting started: Wait till you see what it does to the rest of your brain (in Chapter 9). Yum.

When "Want" Becomes "Need"

The *DSM-V* says all you need for addiction is tolerance and dependence (engaging despite conscious knowledge and recognition of their detriment), with resultant misery. Behaviors and substances that used to be excluded from the definition now qualify under this rubric. Can you honestly look yourself in the mirror and tell yourself that you have no addictions? Ben & Jerry's, eBay, Facebook, porn, video games, coffee? How long did the rush from the new iPhone last? Or the new car? Or the new wife? As a society we've become tolerant by obtaining new stuff at a moment's notice. We don't just *want*, we *need* the newest, fastest, shiniest, classiest, coolest. You might call dopamine the dark underbelly of our consumer culture. It's the driver of desire, the purveyor of pleasure, the neurotransmitter of novelty, the lever that business pushes to keep our economy going, but at a clear, perceptible, and increasing cost. We've purified our substances to concentrate their effects, and we are perpetually in need of the next shiny object.

Apparently that goes for presidents too. The Coolidge effect takes its name from an apocryphal story of an experimental government farm visited by President and Mrs. Coolidge. When Mrs. Coolidge came to the chicken yard she noted a randy rooster. She asked how often the rooster mated and was told, "Dozens of times each day," to which she replied, "Tell that to the president when he comes by." The president asked, "Same hen every time?" "Oh, no, Mr. President, a different hen every time." The president said, "Tell that to Mrs. Coolidge."

PART III

Contentment—
The Bluebird
of Happiness

7.

Contentment
and Serotonin

uestion: Over the course of history, what prescription medication has evidenced the greatest societal impact? Well, you could argue that cholesterol-lowering drugs (statins) are the most prescribed for the treatment and prevention of heart disease, and have made the most money. You could make the case that anti-malarials have saved millions of lives, especially in third-world countries. Protease inhibitors turned AIDS from a ruthless killer into a public nuisance. Non-steroidal anti-inflammatories such as ibuprofen and naproxen have alleviated pain in a majority of people. How about narcotics and anesthetics? Having surgery two hundred years ago without them was extremely unpleasant; although the recent scourge of opiate addiction might be starting to negate its positive impacts. Maybe Viagra? That has certainly increased the happiness of a portion of the population. Pick any of the above. You would be wrong.

Answer: fluoxetine (Prozac). Psychiatric hospitals once were the saddest places on earth (think *One Flew Over the Cuckoo's Nest* [1962]: I still have nightmares about Nurse Ratched), chock-full of patients with schizophrenia (e.g., patients who thought people were out to kill them and/or plotted to kill others, due to dopamine dysfunction) and patients with clinical depression (e.g., people who would have welcomed being killed, due to serotonin dysfunction). But, at its worst, schizophrenia affected only about 1 percent of the population. Consider the fact that major depressive disorder (MDD) affects 16 to 18 percent of the U.S. population at some time in their lives, and that at any given moment 6 to 8 percent of the people you know are affected.[1] This is a very big deal and takes a huge toll on the individual, on his or her family, and on society.

Psychiatric drugs are truly a miracle of Western civilization. For many years, scientists and doctors had been trying to understand what made some people suffer from severe depression while others seemed preternaturally happy and stars of their own Disney movie. In 1952 a serendipitous finding launched the field of modern psychopharmacology. As is the case with the first generation of many mood-stabilizing treatments, we used it to treat a different malady altogether. Patients with tuberculosis (TB) treated with a drug known as isoniazid (INH, still the drug of choice when you are exposed to someone with TB) out of the blue experienced a lifting of their depression. INH worked on the neurotransmitter serotonin (as well as other areas of the brain), and with more trials and focus, scientists were able to pinpoint that it was the effect of serotonin that caused depressed TB patients to reemerge into the world of the living. Thus, scientists learned that serotonin was responsible, in part, for the feelings of happiness and contentment. And, when out of whack, could cause severe irritability and depression.

Plumbing the Depths

There are two kinds of depression. People with "retarded" depression can't get out of bed, and would kill themselves if they had the energy to do so. They often need to be hospitalized to be kept away from themselves. But they pale in numbers compared to the people with "agitated" depression, who are anxious, irritable, sleepless, and just plain miserable. Both types are associated with individuals eating and sleeping either far too much or far too little, both of which are activities that involve serotonin (see Chapters 9 and 18).

When Prozac, the first in the class of selective serotonin reuptake inhibitors (SSRIs), hit the market in 1986, prescriptions for antidepressants shot up a record 400 percent over the next fifteen years.[2] The genius of Prozac was that it didn't matter which form of depression you had. Whether you were climbing the walls or plumbing the depths of your psyche, Prozac could bring you to ground. Figure 7-1 demonstrates how both retarded and agitated depression can be helped by improving serotonin status.

Due to both Reagan's funding cutbacks and Prozac's successes, over the next two decades, in-patient psychiatric facilities closed faster than Blockbuster Video. I watched this phenomenon occur during my medical fellowship: there weren't enough depressed people in the hospitals to keep them open, and nobody cared about the schizophrenics, so they got dumped onto the street, where they remain today.

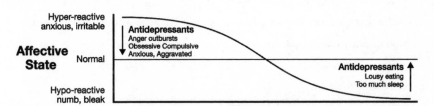

Fig. 7-1: The highs and lows of depression. Depression comes in two varieties—retarded depression (slow thinking and behavior: *I can't get out of bed*); and agitated depression (flight of ideas and inability to concentrate: *I can't get into bed*). SSRIs are antidepressants that, by increasing the amount of serotonin in the synapse, can restore normal levels of mood in either type of depression.

Prozac or one of its many cousins—sertraline (Zoloft), citalopram (Celexa), and paroxetine (Paxil), to name but a few—are now prescribed to alleviate or mitigate a great many mental disorders. Today, SSRIs are the number three most prescribed class of drugs; more people under age sixty-five take antidepressants than any other medication,[3] and as many prescriptions were filled for antidepressants as for cholesterol-lowering drugs.[4] Currently 11 percent of all adolescents are taking an antidepressant[5, 6] not just for depression but for anxiety, anger management, premenstrual syndrome, and obsessive-compulsive disorder as well. The frequency of diagnosis of depression is still on the rise. However, we don't know if this is due to an increased awareness among the medical community (ascertainment bias), if insurance coverage has provided the impetus for overdiagnosing (pills are lucrative to drug companies, and cheaper than psychotherapy), if more adolescents are depressed because of bullying and school pressures, or if people and doctors want to provide a quick fix. But *is* it a fix? Not for everyone. And how does it work?

Before we start talking about serotonin and contentment, let's go back to dopamine and motivation. Dopamine is the reward initiator, and firing of dopamine neurons changes behavior. Remember, the dopamine neurons in the VTA have two primary targets: (1) the nucleus accumbens (NA), where the dopamine signal is translated into desire and reward (I'm stressed, give me a Krispy Kreme), and (2) the prefrontal cortex (PFC), where the dopamine signal is tempered by cognitive control (your personal Jiminy Cricket).

Happy Feet?

But serotonin differs from dopamine in many ways, which makes it difficult to understand and to study. First, serotonin is utilized by different parts of the body. The overwhelming majority (90 percent) is produced and used in the gut, where serotonin is involved in neural and hormonal responses to feeding and how full you are. Another 9 percent can be found in the platelets of our bloodstream, where serotonin helps our blood to clot. That leaves a total of 1 percent of all of your body's serotonin in the brain itself.[7] This is why we can't just measure the amount of serotonin in blood or urine to diagnose depression—because the amount is more a reflection of what's going on in the gut or the bloodstream than in the brain. As an example, carcinoid, which is a tumor of the intestine that overproduces serotonin, causes severe diarrhea, flushing, and abdominal pain and cramping, but it doesn't have very much in the way of central nervous system actions, and it certainly doesn't make its victims happy. But your urine and blood will definitely show high levels of serotonin and its breakdown products (Fig. 7-2). There's no biomarker for depression, no blood test that your doctor can administer. To diagnose clinical depression, doctors use a questionnaire known as the Beck Depression Inventory (BDI), which scores different subjective symptoms of depression. This validated instrument is equivalent to your brain's serotonin meter.

Serotonin neurons fan out to many different part of the brain. When these signals are interpreted either separately or together, we describe the neural experience as some version of happiness. Presumably this is one reason why happiness has so many different definitions, manifestations, and inputs: because different interactions between regions of the brain influence different phenomena—joy, elation, love, etc. We know that serotonin is partially involved in contentment and well-being, but we don't yet have all the details. What's more, dopamine has only five different receptors in the brain (although most of the reward effects are mediated by the D1 and the D2 receptors). In contrast, serotonin has at least fourteen different brain receptors to which it binds, and while there are certain receptors that exert the majority of the serotonin effect, it makes it very difficult to piece together what is happening in any specific brain

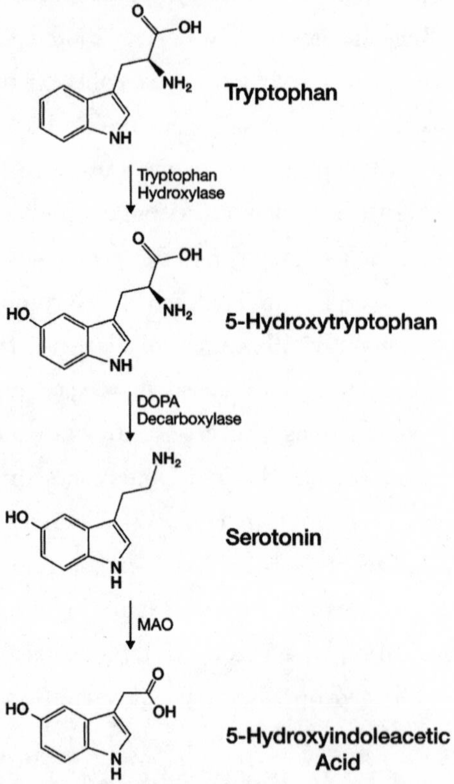

Fig. 7-2: Synthesis and metabolism of serotonin. The amino acid tryptophan receives a hydroxyl group from the enzyme tryptophan hydroxylase to form 5-hydroxytryptophan. This compound is then acted on by DOPA decarboxylase (the same enzyme in the dopamine pathway) to form serotonin. From there, serotonin clearance is achieved by monoamine oxidase.

area. Thus, unlike dopamine, unraveling the role of serotonin in human happiness is a much tougher affair.

Isolating serotonin neurons and figuring out what they do in humans would require some very questionable neurosurgery from some very questionable neurosurgeons (Gene Wilder as Dr. Fronkensteen?). For this reason we have had to primarily use animal models for this work. But this leads to a big question: Is happiness a human attribute exclusively?

How can you tell if an animal is happy? Are there any behaviors that animals demonstrate that are reflections of happiness rather than the result of overlay by reward or pleasure? I've talked to several animal behaviorists at the Society for Behavioral Neuroendocrinology about this. One form of happiness, the nurturing behavior that occurs between parents and offspring, is mediated by oxytocin (the "bonding" hormone) rather than serotonin. But what about general happiness in animals? Ken Locavara, an eminent paleontologist (he discovered the biggest dinosaur remains in Patagonia), suggests that Antarctic penguins repeatedly slide down ice chutes into frozen water, with no secondary gain or reward. There's no food involved, just an expenditure of energy. This behavior can't have any survival advantage—just a general sense of "Wheeeee!!!" So perhaps this is their amusement park and they are demonstrating joy. Or is it pleasure? And that's penguins. Are rats or mice happy? How are we able to tell when a rat or mouse is depressed? For one, we know what they like: sex and sugar. And when they don't perform to get it, they're depressed. Just like us. And we know that antidepressants will alter their behavior. And from the rodent work, we end up extrapolating to humans.

There still exists a large stigma toward the diagnosis and treatment of depression, as if it is a personal moral failing. For many who suffer from depression or have loved ones who do, the idea of it being their fault makes no sense. Who would choose this? Indeed, people with genetic differences anywhere in their brain's serotonin system are at greater risk for suicide.[8] Hardly a choice for these people.

The Sublime Science of Serotonin

Similar to that recounted for dopamine (see Chapter 3), serotonin physiology also has the same three points of regulation. Many things can go wrong, which may cause symptoms of depression. Optimizing each step in the process is necessary to reach our own individual Zen.

(1) Synthesis. Serotonin is an ongoing requirement throughout life. Its primary building block is the amino acid tryptophan, which you must eat—you can't make it. It also happens to be one of the least available items in the human diet. Tryptophan is found in greatest quantity (but still pretty rare) in eggs, fish, and poultry. Many vegetable protein sources are notoriously low in tryptophan. Fewer building blocks means less product: not enough tryptophan in the diet means less serotonin can be made. (More about diet in Chapter 9.)

So, you have a limited amount of tryptophan in your system to make serotonin, which is actually a hot commodity in your brain (Fig. 7-2). Most of the tryptophan consumed is going to be used to produce serotonin in your gut. Only 1 percent is available for your brain. There isn't just one serotonin factory in the body. In fact, once serotonin has been made in the gut or elsewhere, it can't cross the blood-brain barrier. Your brain is on its own, it's got to make serotonin itself. And the brain serotonin factory is localized to a long thin area deep in the most primitive part of the brain, called the raphe nuclei. (We'll focus on the dorsal raphe nucleus, or DRN, from here on. See Fig. 2-1.)

Tryptophan is only one type of amino acid (one of the building blocks of protein) that needs to make it into the brain. These building blocks hop on amino acid transporters to cross over from blood into brain. The problem is, the transporters, like a taxicab at 11:00 p.m. on a snowy New Year's Eve, are sometimes difficult to come by. Tryptophan is in competition with at least two other amino acids, phenylalanine and tyrosine, which are the building blocks for dopamine. So guess what, folks? The more building blocks for dopamine (i.e., reward-seeking behavior) in your blood, the fewer taxis that are available for tryptophan to head to party central in the brain and whip up some contentment for the evening.

This competitive mechanism of tryptophan transport into the brain is but one way by which reward trumps contentment. More are coming (see Chapter 10).

(2) Action. Similar to dopamine, serotonin is released from its nerve terminals and must traverse the synapse to meet up with its receptor. Serotonin nerve terminals are all over the brain in order to bind to different receptors to exert different effects. Thus, the actions of serotonin are much harder to quantify because: (a) there is no clear anatomic location, (b) there are too many receptors to keep track of, and (c) there are many different kinds of responses among people, and even within the same person. Unfortunately, we aren't entirely sure which receptors work which way. For instance, triptans are a class of drugs that bind to two specific serotonin receptors, and they are the best anti-migraine medications that we physicians have at our disposal. But taking these medications does nothing for your state of mind (although if you've ever had a migraine, then not having one is a state of bliss).

One receptor in particular, the serotonin-1a receptor, seems to be uniquely involved in decreasing anxiety and mitigating depression. It's the binding of serotonin to this receptor that is equated to well-being and contentment. We know this because we have been able to genetically remove that specific receptor from mice. When they don't have it, they are extremely anxious and no amount of antidepressant is going to fix it because the receptor is gone.[9] The serotonin-1a receptor has been a hotbed of concern for psychiatric disease for decades.[10] In one Japanese study, genetic serotonin-1a receptor differences are associated with bipolar disorder (formerly called manic-depressive illness).[11] Drugs that bind to the serotonin-1a receptor (known as agonists, or chemical mimics) are a mainstay of antidepressant therapy,[12] and new drugs are coming online at a relatively rapid pace.[13] For instance, buspirone (Buspar) is a commonly used serotonin-1a agonist in the treatment of severe anxiety.

(3) Clearance. After the packets of serotonin transmitters are released from the neuron, they need to traverse the synapse to get to the receptor. After they have bound to the receptor, they hang out in the synapse waiting to be recycled or deactivated. The same process takes place here as it does with dopamine, using the same enzyme monoamine oxidase (MAO), which will degrade serotonin into its waste product 5-hydroxyindole acetic acid (5-HIAA) (Fig. 7-2). The MAO acts as a Pac-Man here as well, essentially gobbling up and destroying serotonin molecules. This is why MAO inhibitors such as phenelzine (Nardil) work as antidepressants, by keeping the levels of serotonin elevated, fostering more chance to bind to a receptor.

Alternatively, the serotonin transporter is a protein that recycles serotonin from the postsynaptic neuron back to the presynaptic neuron so it can be repackaged and used again the next time the neuron fires. These serotonin recyclers/transporters perform the same function as the dopamine transporter mentioned in Chapter 3, acting as "hungry hungry hippos." They will suck the serotonin back into the neuron to be recycled and released again. This is the site of action of all the newer selective serotonin reuptake inhibitors (SSRIs), like fluoxetine (Prozac), sertraline (Zoloft), citalopram (Celexa), and escitalopram (Lexapro) to increase the amount of serotonin within the synapse in order to elevate mood. So what these SSRIs do is basically put a muzzle on the hungry hungry hippos. They are still functional, just less so. However, you don't want to knock them out of the game completely. Having too much serotonin in the synapse can also be a problem (read on).

Always Look on the Bright Side of Life

How well your serotonin recycler/transporter works has a lot to do with how happy you are. Temperament goes a long way in explaining happiness, and differences in the serotonin transporter go a long way in explaining differences in temperament.[14] For instance, those born with a specific allele (genetic variation) of their serotonin transporter (the 5-HTTLPR) are quite anxious as children, and are more likely to suffer into adulthood as a result of an unstable home life[15] (i.e., have a greater propensity for anxiety, depression, and drug abuse[16]).

As an interesting aside, despite consistently experiencing more adverse circumstances throughout American society, African-American adults routinely exhibit a lower incidence of clinical depression than do Caucasians and Latinos.[17] This is not explainable by sampling differences, sex differences, or levels of education. African-Americans exhibit less anxiety than Caucasians do.[18] There may be several reasons for this dichotomy. One thought is that these questionnaires may be culturally biased, which may in fact be true. Another possibility is that the African-American population of the U.S. exhibits a higher affiliation with a religious denomination than any other racial group,[19] which may provide them with a social basis for achieving happiness despite socioeconomic adversity. But there may be a biochemical reason as well. African-Americans are known to have a genetic difference in the gene that encodes the 5-HTTLPR (serotonin transporter, aka hungry hungry hippos), which reduces the ability to clear serotonin from the synapse.[20] Thus, African-Americans may have their own built-in SSRI, so they get less depressed in the face of adverse circumstances.

But just as with dopamine, too much of a good thing can become a bad thing. Serotonin can have serious side effects, including irritability and suicidal thoughts and actions.[21] Excessive serotonin effects can lead to negative levels of mood, and outward behaviors such as impulsive aggression, because of binding to receptors other than the -1a receptor.[22] Serotonin syndrome, which results from too much serotonin activity because of SSRI overdose or interactions with other drugs, is characterized by changes in mental state and muscle tone, and autonomic nervous system problems.[23] Going overboard on serotonin can take someone who's morose and give them just enough brain activity and mental energy to make them suicidal, which is why people on antidepressants shouldn't dose themselves. Just as with dopamine, the goal is not to increase your serotonin status indiscriminately but rather to find your sweet spot.

Most of us, as numerous surveys indicate, concede that the most important goal of life is happiness. But the quest for happiness begins and ends with optimization of your serotonin neurotransmission[24]—clearly no easy feat. Chances are you've seen a commercial for an antidepressant. It generally starts with a woman looking forlorn and in cold, gray weather. Then, magically, the sun is shining, she is smiling, her kids are well behaved, and they all live happily ever after.

Alas, there is no one magic pill. Medications work differently on different people and they may lose or gain effectiveness over the life span. The dosage of Prozac taken by an eighteen-year-old may not work the same way when he or she is forty. After giving birth, women's hormones go into a tailspin and they may experience postpartum depression, necessitating the usage of antidepressants. After a year or so, their serotonin may go back to normal on its own, or it might not. Anti-depressant medications can work wonders, but only 25 percent of those who take them experience a full remission.[25] The remaining 75 percent may experience some relief but not complete reversal of symptoms. More aid is needed. Even for those of us who do not suffer from depression, few of us know how to attain contentment. Short of SSRIs, what hope do we have of achieving any meaningful happiness in this life?

Are we really Prozac Nation? Not quite. Read on.

8.

Picking the Lock
to Nirvana

I f you're afloat, peering up at a panorama of tangerine trees and marmalade skies, lower your gaze. Chances are your shipmate just might
be a girl with kaleidoscope eyes.

John Lennon was one of the chief spokespeople of the counterculture of
the 1960s. "Lucy in the Sky with Diamonds" (1967), composed by Lennon
and Paul McCartney, extolled the benefits of the synthetic psychedelic lysergic acid diethylamide (LSD), and the growing desire of young people to,
in the words of Harvard psychiatrist, political activist, and eventual public
enemy number one, Dr. Timothy Leary, "Turn on, tune in, and drop out."

LSD was first manufactured in a Swiss lab by pharmaceutical chemist
Albert Hofmann in 1938, but first ingested by Hofmann in 1943.[1] Immediately, scientists and researchers saw its potential—it was used in attempts to cure autism and treat convicts, among others—and the first
commercial preparation, Delysid, hit the European market in 1947. Many
different mind-altering drugs entered our societal lexicon during this

period. Mescaline, a phenylethylamine derivative used in traditional Native American worship rituals, was purified from the peyote cactus. Psilocin, the active form of the tryptamine precursor psilocybin, was purified from indigenous "magic mushrooms" found in Mexico. While new to the American mainstream, these plants had been used for hundreds, sometimes even thousands, of years by different indigenous groups and cultures. Rituals involving naturally occurring hallucinogens have played a central role in the religions, and sometimes even the language of various tribes—in quests to find spirit animals, communicate with the dead, and seek out the divine. When Hofmann created LSD, all of a sudden scientists wanted in.

Drinking the Electric Kool-Aid

In 1953 the structure of serotonin and its presence in the brain was confirmed.[2] Scientists soon thereafter discovered the incredible structural similarities between serotonin and some of these compounds—especially psilocybin and LSD (Fig. 8-1). Thus began a seventeen-year scientific and

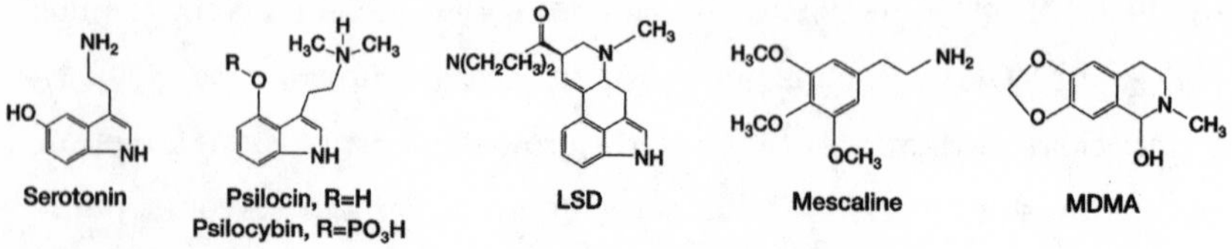

Serotonin   Psilocin, R=H   LSD   Mescaline   MDMA
                   Psilocybin, R=PO$_3$H

Fig. 8-1: Serotonin receptor "skeleton keys." Psychedelics are modifications of the structure of the parent compound serotonin. These changes allow different compounds to bind selectively to individual serotonin receptors instead of all sixteen. But some still cross-react. The tryptamine derivatives psilocybin and LSD can bind to both the serotonin-2a receptor (the mystical experience) and the serotonin-1a receptor (contentment). The phenylethylamine compound mescaline binds only to the serotonin-2a receptor. MDMA, or ecstasy (see Chapter 10), not only binds to the serotonin-2a receptor, it binds to the dopamine receptor as well.

existential quest to unravel the hidden mysteries of the mind, and in particular, the quest for happiness—both natural and artificial. One set of scientists started altering the molecular structure of these compounds to increase their potency, while another set of scientists labeled them with radioactivity to look at their binding sites in the brain and their mechanisms of action. After years of trial and error, they discerned that these compounds acted as a serotonin agonist, meaning that they mimicked serotonin and would bind to specific serotonin receptors in the brain; namely, the -1a (see Chapter 7) and the -2a receptors.

The 1960s was the golden age of LSD research. The U.S. government subsidized at least 116 experiments (that we know of) over this interval to unlock its secrets. Dr. Stanislav Grof, one of the early experimenters, described LSD as a "non-specific amplifier of the unconscious,"[3] for both good and bad. The suggestion was that LSD might be a primary modulator of the unconscious mind, and unlocking its mysteries would answer the questions of who we are, why we are here, and what's to become of us. Big questions indeed. Maybe too big to be left to scientists?

As hard as you may try, you can't keep something this big locked up in the lab. These molecules escaped from the ivory tower and started a (relatively) bloodless revolution within America, especially among young people, who were disillusioned with the U.S. government, and the handling of the Vietnam War and the civil rights movement. Psychedelics were all the rage in the late 1960s throughout the country. College campuses were the testing ground for this social experiment, and some still are.

Three observations about the use and users of psychedelics should be made at this point.

1. Some users of psychedelics would experience "bad trips"; that is, they would experience unwanted fear and paranoia. Hallucinogenic experiences can't be easily predicted. Maybe someone will

have a good/mellow trip, feel at one with the universe, and talk to the deities—or maybe they will feel that their face is melting off and the world is contracting. It's hard to predict what, who, and how these drugs cause a bad trip. In general, hallucinogens magnify the emotional and mental state of the user at the time. If someone is depressed or manic, a hallucinogen, taken on its own, would likely intensify the feeling in the same direction. Based on anecdotal data, the psychedelic experience is, in the words of Timothy Leary, responsive to both "set" (i.e., mind-set) and "setting" (i.e., place and people you are with). Perhaps this was best typified by the inconsistent and incoherent results of a clandestine CIA operation called the MK-ULTRA program (aka Operation Midnight Climax), which between 1953 and 1964 dosed unsuspecting military personnel and unwitting victims in New York and San Francisco with LSD in their alcoholic drinks.[4] Ostensibly, the reason for this covert program was that the CIA was concerned that Russia, Communist China, and North Korea were using these drugs to brainwash American prisoners of war—think Laurence Harvey in *The Manchurian Candidate* (1962) (Queen of Diamonds, anyone?)—and they needed to fight back. The responses of these "volunteers" ranged from anxiety to sheer paranoia to apparent psychosis: their world did not make sense, because they were navigating blind. Thus, the need for informed consent and a tour guide for your metaphysical trip.

2. Although some of these compounds demonstrated decreased efficacy (i.e., tolerance) with repeated use, few users of psychedelics demonstrated either dependence or withdrawal upon quitting.[5] Most were able to walk away from their use without untoward personal or societal consequences. Virtually no emergency room visits, no spike in crime, and no users rushed into rehab, as is often the case when dopamine agonists (e.g., cocaine) or opiates (e.g.,

heroin) are withdrawn. It is estimated that as many as 30 million people worldwide have come into contact with a psychedelic drug at some point in time.[6] For instance, a recent examination of the National Survey of Drug Use and Health demonstrated that of the respondents, 13.4 percent admitted to long-term psychedelic use. Yet, despite chronic use, these same people reported no drug addiction, and surprisingly little coincident mental illness; in fact, the prevalence of psychiatric diagnoses was lower in these users than in the general population,[7] and few ended up in mental institutions (with the exception of some who have metaphorically fried their brains from too much acid). In other words, psychedelics are not classically addictive.

3. A third and perhaps more important observation: recent research indicates that when LSD is ingested in a controlled setting, long-lasting effects are sometimes experienced, which can include improved social relationships with family, increased physical and psychological self-care, and increased sense of spirituality.[8] Whether these feelings reflect a biochemical change in the brain, or are just an uplifting sequel to the mystical experience is unknown. But people report, for lack of a better term, mind-altering aftereffects.

The Feds Raid the Party

Proponents of psychedelic use such as Leary were gaining a foothold with America's youth in the 1960s; it was a message that conflated with their anti-war sentiments. To quell the movement, California state senator Donald Grunsky introduced a law into the state legislature that banned possession, distribution, and importation of LSD and its cousin dimethyltryptamine (DMT), which was signed by Governor Ronald Reagan in 1966. The backlash culminated in the pinnacle of American counterculture: San Francisco's 1967 Summer of Love, the vestiges of which are still apparent on Haight Street (come visit!—just ignore the hypodermic needles). Indeed, America's youth, the first wave of the baby boomer generation, was dropping out in droves. "Think for yourself, and question authority" was Leary's motto. Throughout adolescence and early adulthood, the cognitive connections between actions and consequences are muddled, as the maturation of the prefrontal cortex (the Jiminy Cricket) is not complete until approximately twenty-five years of age.[9] (This is also why the actuaries jack up auto insurance rates until you reach your twenty-fifth birthday.) The baby boomers who attended Woodstock in 1969 struck fear in the heart of the U.S. government. After all, the Army needs young men to fight in wars. Generals and admirals in the armed forces, witnessing a clear change in young men's taste for participating in armed conflict, advised the Nixon administration that these compounds were among the most dangerous and destructive drugs ever devised—even more destructive than opiates.

The counterculture movement abruptly went underground with Congress's passage of the Controlled Substances Enforcement Act of 1970 and the establishment of the Drug Enforcement Administration in 1973, which was charged with regulating all dopamine, opioid, cannabinoid, and serotonin agonists. Heroin, marijuana, and all psychedelics were thereafter classified as Schedule I, meaning that they had no medicinal importance and no legal purpose; in other words, banned. With the stroke of a pen, Richard Nixon wiped out a fascinating and potentially promising line of medical and psychiatric research. Now relegated to the dustbin of scientific history, this work would languish for the next forty years. Deleted from our collective memory is the fact that some of the users of these compounds experienced what, for lack of a better phrase, was a "life transformation." The anthem for this movement, Lennon's *Imagine*, told young people to lay down their guns, part with their worldly possessions, and "learn to live as one." Why did he believe this? Because he was singing Kumbaya with Lucy in the Sky with Diamonds? Can hallucinogens make you happy or, at a minimum, content? Not always: some have reported disembodiment and severe anxiety. And for how long? The length of the drug trip itself (which, for LSD, can be a very long twelve-plus hours)? Or are there lasting effects? Days? Months? Are humans happier in an altered state? The secrets of life, love, happiness, and contentment were buried in a tomb too dangerous to excavate.

Fast-forward to today. A courageous group of doctors and scientists are excavating that tomb right now. Some of these drugs are making a resurgence in science, albeit under extremely strict government oversight. Michael Pollan's article "The Trip Treatment" in the *New Yorker*[10] recounts the human interest story behind how these drugs have been rediscovered by some forward-thinking clinicians with the help of power brokers invested in unlocking their mysteries.

A New Death with Dignity?

Who on this earth is in greatest need of happiness, or at least the alleviation of the severest form of dysphoria or distress? Terminal cancer patients, that's who. Standard hospice care provides such patients with opiates like hydromorphone (Dilaudid), which, while alleviating pain, dope them up to the point where they can't and don't care, and can't even respond: they can't tell their doctors that they are scared, or their loved ones that they love them. And of course these opiates are highly addictive. You could argue: Who cares about addiction if you're already dying? Both of my parents died in hospice care, both doped up on opiates at the end. I couldn't tell them I loved them, and they couldn't communicate back. Prescribing opiates is more humane than letting patients suffer but nonetheless not an optimal way to depart this world. We *all* deserve a better exit than that, at peace with our own imminent mortality.

In a study that took a full decade to complete, and with the approval of the FDA, NIH, DEA, and a host of institutional review boards, Charles Grob at Harbor-UCLA Medical Center assessed the use of psilocybin (the compound in "magic mushrooms") as a stand-alone treatment for the reactive anxiety and depression that attends death due to terminal cancer. In an initial study, twelve individuals with a life-threatening cancer diagnosis participated in a double-blind randomized crossover fashion (neither the subject nor the physician knew which treatment was being administered) with either psilocybin or niacin (Vitamin B$_3$), which results in a tingling sensation, and acted as the placebo control.[11] Furthermore, every subject was prepared by a licensed psychologist beforehand to minimize the possibility of any side effects or a bad trip. Each had their own personalized metaphysical tour guide, who remained with them through the session. They optimized the set and the setting by providing a pleasing and comfortable environment. These clinical research studies were carefully performed and documented, and above reproach. The results were quite remarkable. Feelings of "oceanic boundlessness" and "visionary restructuralization" were followed by positive mood and reduction in depressive scores, which persisted up to six months after the psilocybin treatment ended.

Several follow-up studies are now being conducted. Stephen Ross at NYU School of Medicine randomized twenty-nine participants with cancer in a double-blind fashion to receive either psilocybin or niacin. Again, reductions in long-term anxiety and depression were observed, and with long-lasting effects still measurable six months after hallucinogen exposure; and again the benefit correlated with the extent of the "mystical experience."[12] Using LSD as the hallucinogen, Peter Gasser in Switzerland[13] showed that twelve cancer patients also showed short- and long-term benefit, and with no persistent side effects beyond the day of the study itself. Further studies have corroborated these beneficial effects up to fourteen months out.[14]

Due to the remarkable long-term nature of these clinical responses and the lack of long-term side effects, such studies are expanding. Currently, clinicians throughout the world are testing whether these compounds can treat addictions such as tobacco and alcohol.[15] What? A psychedelic drug can confer contentment, even if artificial, or can reverse long-standing substance abuse? Not so fast. We'll deal with this in Chapter 10.

Special on Receptors—
Buy One, Get One Free!

Clearly, hallucinations and contentment are not the same thing. You don't have to be in a mind-altered state to be happy. Second, most happy people have not lost touch with reality. Lastly and most importantly, not everyone who has experienced the effects of psychedelics has been moved to give up all their earthly possessions and live in a yurt. Nonetheless, there is clearly some form of overlap. What ties serotonin, hallucinations, and contentment together?

Although I can't prove this, the key to this puzzle may very well lie in the nature of the compounds themselves, which serotonin receptors they activate, and where and how much it takes to activate them. Remember from Chapter 7, while originating in the DRN (one area of the midbrain), serotonin acts throughout the cerebral cortex, where it binds to as many as fourteen different receptors coded by eighteen different genes, likely mediating serotonin's various cognitive, behavioral, and experiential effects. We also recall that the primary modulator of serotonin's effects are the SSRIs, which improve mood and alleviate depression. As far as we can tell, the effects of SSRIs on anxiety and depression are on the serotonin transporter located at the serotonin-1a receptor.[16, 17] How do we know? Because if you knock out that specific receptor in mice, they become incredibly anxious, and SSRIs can't rescue them,[18] yet knockout of

other receptors doesn't lead to depression. And because genetic polymorphisms in the seroronin-1a receptor predispose humans to major depressive disorder,[19] the -1a receptor appears to be the seat of our contentment
and well-being.

In contrast, through painstaking experiments on animals and humans,
the mind-altering effects of all psychedelic compounds have been traced
to their stimulatory effects on the serotonin-2a receptor. Not -1a but -2a.
Punch your tickets to the Magical Mystery Tour—they are all chauffeured
by the -2a receptor, whether it's snorting Yopo, dropping acid, downing
ayahuasca, or licking the Colorado River toad (yes, really—and do not try
this at home).

Where are these serotonin-2a receptors that generate these vivid and
otherworldly effects? Recently, Robin Carhart-Harris of Imperial College London delineated the two primary brain sites of hallucinogen action.[20] First, the visual cortex. Injection of radio-labeled psilocybin lights
up the visual cortex like a Christmas tree.[21] Perhaps this is not all that
surprising, given Lennon's experiences of tangerine trees and marmalade
skies. They are also in the prefrontal cortex (our Jiminy Cricket), which
may explain why these compounds alter our inhibitions and increase sensations of reward.

But action on serotonin-2a receptors doesn't explain the connection between some of the psychedelics and long-term contentment. You'd think
that if all the hallucinogenic drugs bound to and activated the same sero-
tonin-2a receptor, they would act in the same fashion and exert the same
effects. But not all do. The phenylethylamine class of compounds, of which
mescaline is the natural version (Fig. 8-1), isn't associated with the post-
administration experience of contentment.[22] Rather, the afterglow appears
to be restricted to the tryptamine class of compounds, of which psilocybin
is the natural version.[23] In fact, the drugs that provide contentment (-1a
binding) on top of the mystical experiences (-2a binding) are all of the
tryptamine class; and you have to reach a dose that achieves the mystical

experience in order to experience the post-dosing contentment.[24] Hey, two receptors for the price of one![25] In fact, virtually the entire tryptamine class of psychedelics (to which psilocybin and LSD belong) bind to *both* the -1a and -2a receptors.[26] In contrast, mescaline binds just fine to the -2a receptor to provide the hallucinogenic experience, but it has little effect on the -1a receptor,[27, 28] which likely accounts for the lack of the afterglow.

Could this added receptor bonus really explain the ability of psilocybin to remove angst and fear from terminal cancer patients? Can this extra effect really treat alcohol and tobacco addiction? Can this class of drugs really cause lions to lie down with lambs? Doubt it. I have met quite a few people who have dabbled in taking hallucinogens and none of them have become monks, although some did move to Marin County. Many addicts have at some time taken LSD and are still addicted to their drug of choice. Does the tour guide make a difference? Or the dosage? We don't know yet . . . but can you see why the armed forces were so scared of the fallout?

The Psychedelic Hangover

More recent well-controlled studies of LSD administration in normal non-depressed volunteers suggest that the drug induces profound perceptual changes: the way these subjects see the world around them. Volunteers scored significantly higher on the Creative Imagination Scale,[29] and exhibited more openness to new ideas and new experiences.[30] Steve Jobs swore by LSD—until he started Apple. Then it became a distraction. What would it do to you? Chances are you won't start a company, but you might end up down a rabbit hole that you can't climb out of. You're taking your brain in your hands. You ready for that?

Yet the implications of this research are nothing short of life- and world-altering. Fifty years ago it was a free-for-all. Then the pendulum swung in the opposite direction and the feds raided the party. Is there a happy medium? The pendulum is just now starting to swing back. What if tryptamine psychedelics (LSD and psilocybin) were reclassified to Schedule II, where doctors could prescribe them to selected patients in controlled settings? What if the taboo of hallucinogens was removed and we had "medical mushrooms"?

The big issue with all centrally acting drugs is the concern over tolerance and either withdrawal or dependence—in other words, their addictive potential (see Chapter 5). Despite demonstrating tolerance, these serotonin agonists have rarely been shown to lead to withdrawal or dependence; in other words, they do not appear to be classically addictive. In fact, they are now being evaluated to treat addictions to other drugs![31] Serotonin affects dopamine? We're going there in Chapter 10.

Nonetheless, these serotonin agonists are not completely safe. High doses can on rare occasions cause constriction of the blood vessels and coronary artery spasms, so their recreational use without a doctor's supervision is contraindicated. There is no doubt that repeated daily dosing of LSD leads to reduction of effect[32] due to down-regulation of serotonin-2a receptors,[33] which might have long-term sequelae that we just don't know about. And the bad trips? This was ostensibly the reason Congress banned psychedelics back in 1970. In drug parlance, we're talking about a very narrow therapeutic window, and if you're not within the window, you might just jump out of it instead. Some of the newer designer hallucinogens can still elicit the occasional bout of agitation, rapid heartbeat, sweating, and combativeness that requires an ER visit and IV sedation until the drug wears off. Not ready for prime time, to say the least.

Better Living Through Biochemistry?

These studies provide yet another line of reasoning to support the assertion that I am trying to drive home—that our emotions are just the inward expression of biochemical processes in the brain. In the case of hallucinogens, signaling of the serotonin-1a receptor drives contentment, whereas signaling of the serotonin-2a receptor drives the mystical experience. In our modern society the role of mind-altering drugs to achieve heightened consciousness and/or contentment has yet to be determined, and will require careful scientific investigation in controlled settings along with philosophical and ethical debate before the public can be trusted with the key to nirvana.

We *are* our biochemistry, whether we like it or not. And our biochemistry can be manipulated. Sometimes naturally and sometimes artificially. Sometimes by ourselves but sometimes by others. Sometimes for good and sometimes for ill.

9.

What You Eat in Private You Wear in Public

t would seem that optimizing serotonin availability and action at the -1a receptor is the neurochemical key to contentment. Indeed, it is. But producing adequate amounts of brain serotonin all the time is hard to do and even harder to maintain. No wonder many of us are unhappy. Our current drug armamentarium is pretty sparse, and fraught with tendencies to overshoot along with the mental health dangers that come with it—everything from severe irritability to bad trips to suicide. And these many medications—from your doc, shaman, or college roommate—are not actually making serotonin. Some antidepressants and hallucinogens are acting as serotonin agonists (chemical mimics). The natural stuff is up to your brain to make. And for that you need the basic ingredients or building blocks. For serotonin, that would be tryptophan, the basic ingredient to inner contentment. But our food supply isn't doing us any favors either, given that:

1. the precursor amino acid tryptophan is the rarest amino acid in our diet;

2. ingested tryptophan can be metabolized by an alternate pathway in the body (called kyurenine), which can create inflammatory by-products, leading to worsening of your health;

3. 99 percent of the ingested tryptophan is converted to serotonin in other parts of the body (used for other functions in your gut and blood) before it even gets a chance to be transported into the brain;

4. tryptophan transport from the blood into the brain is in competition with much higher concentrations of both tyrosine and phenylalanine (the precursors for dopamine) for the same blood-brain transporter;

5. the serotonin that is manufactured from tryptophan elsewhere in the body can't gain access to the brain, because it can't cross the blood-brain barrier;

6. the serotonin receptors are virtually everywhere throughout the brain, so there's a big demand for that small supply; and

7. the enzyme that inactivates serotonin (MAO, the ubiquitous Pac-Man) is very good at its job.

It's no wonder we're unhappy: we're always playing catch-up. Most of us are functionally serotonin-deficient some of the time (a state), and some of us are serotonin-deficient all the time (a trait), both of which can influence clinical depression. There's barely enough serotonin in the brain to generate even a fleeting feeling of contentment in the first place.

Eating for Sleeping

Serotonin is known to prepare you for sleep[1] and influence sleep stage cycles, especially by reducing the active or rapid eye movement (REM) phase and increasing the slow wave or inactive phase of sleep. If you give tryptophan (the precursor to serotonin) to normal non-depressed people, you tend to see some lethargy and sleepiness and decreased reaction time;[2] perhaps this is one of the reasons people blame the tryptophan for falling asleep at their Thanksgiving turkey dinner (besides stuffing yourself with stuffing, mashed potato, pumpkin, and pecan pie so that you can't move). Conversely, if you provide adults with a tryptophan-depleting drink just prior to bedtime to reduce their brain serotonin, their sleep looks like people who have untreated clinical depression, which keeps them awake at night.[3] SSRIs, while they treat symptoms of depression, also push the reset button on your sleep cycle, often with the result that you can't sleep or you sleep too much. Thus, the tryptophan in your diet goes a long way toward determining how well you sleep, and how well you sleep goes a long way in determining your level of contentment. Our increased levels of sugar and caffeine in the diet sure don't help matters any. We're wired from Red Bull and Starbucks Frappucinos, making us even less likely to get regular sleep and wreaking havoc on our metabolic systems in general. For instance, a study of fourth and seventh graders shows a correlation between shortened sleep and soda consumption,[4] although we can't determine if this is due to the sugar or the caffeine. More on the role of sleep in unhappiness in Chapter 10.

While there are several steps involved in converting the tryptophan in food into serotonin within the brain, most of us don't get enough dietary tryptophan in the first place. Eggs and fish have high concentrations of tryptophan, nuts and poultry are not too far behind,[5] and spinach and soy make the list as well. Although beware: just because they are advertised as Chicken McNuggets doesn't mean there's any chicken in them.[6] Indeed, people who are egg eaters and fish eaters have the highest levels of tryptophan consumption as measured by blood concentrations (of course, looking at levels in the blood doesn't mean that any tryptophan actually made it into the brain, so it's by no means a perfect marker).[7,8] Fish consumption is inversely related to depression in large meta-analyses,[9] although we don't yet have cause and effect. But eggs are frequently omitted from certain processed foods because they curdle with time, because they can go rancid when not refrigerated or when they're old, and because enough people are allergic to eggs. And fish is not usually a big seller as an ingredient in processed food, in part because certain fish don't freeze well and most people want to see the fish to determine how fresh it is.

The nutraceutical industry is actually peddling both tryptophan as well as the next-step chemical intermediate on the way to making serotonin (5-hydroxytryptophan) (see Fig. 7-2) in capsule form. Currently randomized placebo-controlled trials of these nutraceuticals to improve depression are early and limited.[10,11] One group performed a double-blind placebo-controlled trial giving tryptophan to a bunch of petulant people, and lo and behold, they got nicer.[12] (Maybe slip some tryptophan in your boss's coffee tomorrow . . .) A meta-analysis does argue for some benefit in depressed patients,[13] with some attendant side effects. However, Big Pharma isn't interested in going down this road, because a tryptophan pill isn't patentable, and they can sell SSRIs and charge a bundle. We really don't know what a tryptophan-replete America would look like.

What's Your Beef?

What about red meat, which is supposed to provide high-quality protein? America prides itself on its meat production, and its meat consumption. Does red meat have enough tryptophan? It does, but let's have a look at the difference between the corn-fed beef of processed food versus beef that came from cattle that were grass-fed. Turns out that corn is relatively deficient in tryptophan but is loaded with phenylalanine and tyrosine, the precursors of dopamine.[14] While contested by the processed food industry, it is likely that we who eat these animals are not getting very much tryptophan. Furthermore, corn-fed beef has higher levels of branched-chain amino acids (leucine, isoleucine, valine) that contribute to liver fat, which drives the *metabolic syndrome* (see below). Chicken is the one processed food staple that contains a reasonably high quantity of tryptophan,[15] but there's a big catch. Chickens raised for the processed food industry are corn-fed, just like the cattle, and contain a lot of branched-chain amino acids as well.

Each of these amino acids is "essential," which means you have to eat them; your body can't synthesize them. Branched-chain amino acids account for over 20 percent of all the amino acids (building blocks) in the Western diet.[16] If you're in puberty, or a bodybuilder, then these branched-chain amino acids are necessary for building the proteins that are found in muscle (that's what's in protein powder). But if you're a mere mortal like me and most of the sedentary world, then chowing down on excess branched-chain amino acids means there's no place to store them, which means that the liver has more of these amino acids to process and metabolize into energy. The energy overload in the liver drives fat accumulation and insulin resistance,[17] promoting all the chronic metabolic diseases that are part of metabolic syndrome,[18] which will directly affect both your physical and mental health.

Metabolic syndrome is the smorgasbord of chronic metabolic diseases from which America, and indeed the entire world, now suffers. To name them: how about heart disease, hypertension, blood lipid problems such as hypertriglyceridemia, type 2 diabetes, non-alcoholic fatty liver disease, chronic kidney disease, polycystic ovarian disease, cancer, and dementia? These are the diseases of insulin resistance, where insulin doesn't clear glucose from the blood properly, while fat precipitates in your liver and muscles. Next time you're at the butcher, have him show you strip steak from a grass-fed cow and a corn-fed cow. The grass-fed steak is pink and pretty homogeneous throughout. It's delicious, but when you grill it up, it's a little tough. Now look at the corn-fed steak. See all that marbling?

We love it, because that's where the flavor is. And after grilling, it practically cuts with a butter knife. That marbling is fat in the muscle. That's muscle insulin resistance. That cow had metabolic syndrome; we just happened to slaughter it before it got sick, and now we're consuming the aftereffects in each and every Big Mac.

What Am I, Chopped Liver?

At the cellular level, the avalanche of energy from a processed food meal overwhelms your liver's cellular power generators—the mitochondria. When these liver mitochondria get overloaded, they have no choice but to turn the extra energy into liver fat. These molecules of liver fat have one of two fates: either (1) your liver can package them into very-low-density lipoproteins (VLDL), which can lead to heart disease and obesity, and which your doctor can measure as serum triglycerides on your lab panel; or (2) your liver can't package them, they turn into fat droplets, and they make your liver sick. A sick liver doesn't respond well to insulin, causing the pancreas to release excess insulin. Eventually your pancreas gives out, and now you have type 2 diabetes. You've got metabolic syndrome and are losing years of life as your cells and your body age more quickly. How can all this be happening when you're dieting and buying low-fat products? That's just what causes it! That's what's happened to America, and the world. We lowered the fat and put in more sugar to make our food palatable. And all of that increased sugar in the diet is a leading contributing factor to metabolic syndrome. I wrote a whole book about it (*Fat Chance*).

People with metabolic syndrome have decreased serotonin function[19] and are at very high risk for depression. And not because they're fat: thin people get metabolic syndrome also,[20] with the underlying phenomena of insulin resistance and liver fat;[21] they're called TOFI, or "thin on the outside, fat on the inside." Furthermore, each of the disease components (e.g., lipid problems, glucose intolerance) correlates with depression even better than abdominal obesity does.[22] The foods that drive metabolic syndrome are those that are most clearly associated with the foods that people with binge-eating disorder consume with the greatest avidity: refined carbs and sugar.[23] The question is, does the depression drive the food choices, which then drive the metabolic syndrome? Or do the food choices drive the metabolic syndrome that then drives the depression? Which is cause and which is effect? We still don't know. But what we do know is that some people can eat their way out of both the metabolic disease[24] *and* out of the depression[25] by switching to a Mediterranean diet—and why not? Lots of eggs, fish, nuts, fiber, and not very much refined carbohydrate or sugar. The fact that your food choices can lift your mood certainly argues that the food is the driver.

The Slippery Slope

Not only is metabolic syndrome related to depression, it is also related to cognitive decline—and nothing will make you more depressed than losing your intelligence. We've known for a long time that people with type 2 diabetes demonstrate cognitive decline[26] and that brain insulin resistance correlates with dementia (e.g., Alzheimer's disease).[27] Yet type 1 diabetes—which is a spontaneously occurring disease as opposed to type 2, which is driven in part by the sugar in your diet—doesn't correlate with dementia. Both forms of diabetes share high blood glucose as the primary phenomenon. But type 1 diabetics are deficient in insulin, while type 2 diabetics have too much insulin but are resistant to its actions. Because it's not the glucose! It's the insulin! The insulin resistance, driven by all that excess soda and sugar-laden foods, is what leads to the brain plaques that define Alzheimer's disease.[28]

When you're healthy and insulin-sensitive, the insulin rise from a meal normally tells the brain you've had enough food. But once metabolic syndrome sets in, the chronic insulin in the brain does exactly the opposite: it blocks the signal to stop eating.[29,30] Worse yet, the insulin resistance alone (unrelated to blood glucose) predicts cognitive impairment[31] and risk for Alzheimer's disease.[32] It's always been assumed that dementia and cognitive decline are the province of the aged. Not so. Antonio Convit at NYU School of Medicine has shown that teenagers with metabolic syndrome (matched by age, socioeconomic level, school grade, gender, and ethnicity) manifest cognitive decline, brain shrinkage, and decreased white matter integrity.[33] And these kids don't even have type 2 diabetes yet! But they will. In fact, this is a positive feedback cycle. The more brain insulin resistance they develop, the more their dopamine neurons fire, the less restraint of the reward system (your Jiminy Cricket from Chapter 4), the more their anxiety, the less their cognitive inhibition, the more food (especially sugar) they consume, and the more insulin resistance they will develop. Ultimately this vicious cycle culminates in diabetes, dementia,[34] and often depression. Indeed, dietary sugar (sucrose, consisting of glucose and fructose)—rather than blood sugar (glucose)—is the driver in this scenario, because dietary sugar underlies insulin resistance, and insulin resistance underlies dementia.[35] If you feed fructose to animals, you get all the pathology and cognitive decline one sees in Alzheimer's disease,[36] and it causes changes in genes that predict Alzheimer's.[37] So far in humans, we only have correlation. For instance, sugar consumption correlates with risk for dementia in epidemiologic studies.[38,39] But correlation is not causation. In humans, we still don't have cause and effect. Does sugar consumption make you lose brain cells, including those that house your serotonin receptors? Or does losing brain cells make you consume more sugar? Both, most likely. But today we just can't say for sure. Nonetheless, are you comfortable with the risk?

By driving reward, sugar drives risk for addiction (see Chapters 5 and

6), and addiction culminates in unhappiness. Both high-glycemic-index diets (i.e., highly refined carbohydrate from processed foods) and high-added-sugar diets are correlated with depression,[40, 41, 42] but again, correlation is not causation. Does sugar cause depression? Or do depressed people consume sugar to give themselves what little pleasure they can muster? There is no doubt that Cathy Guisewite's eponymous cartoon character Cathy is both depressed and a chocoholic. But are these related, and which caused which? Is sugar consumption a contributor of the depressed state? Or is addiction a necessary intermediate step between sugar consumption and depression? The answer to all of these is a definite maybe.

Your Gut Feeling

But the more immediate question for all of us: Is your sugar consumption really under your control? Common wisdom says you are in control of every item you put in your mouth. And common sense would also suggest that is true (except we know from Part II that in addiction, you're really not in control—the drugs are doing the talking via dopamine). What if something else is also talking, feeding your brain with devious and distracting thoughts—like your bacteria? You might think that your gut microbiome, consisting of 100 trillion bacteria representing several hundred species and residing inside your intestine, would not be immediately connected to your brain. Nonetheless, your gut microbiome appears to have a mind of its own, and it very well may control yours.[43, 44]

Every person on earth harbors his or her own individual Amazon rain forest, with unique creatures living on and in them. The bacterial species found within a person's gut can identify him or her by a unique microbial signature.[45] Yet the human gut microbiome is very clearly and rapidly responsive (within as little as two days) to dietary manipulation,[46] and why not? Different nutrients will make their way farther down the intestine based on different aspects of the diet, such as what carbohydrates you are consuming, whether those carbohydrates are fermented, and whether they are accompanied by the corresponding fiber inherent in that food.[47] Different bacteria like to grow in different dietary "soups" to different extents. In fact, changes in the microbiome have been associated with increased risk for obesity in both directions. For instance, transferring obesogenic bacteria from one mouse strain to another can cause the recipient mouse to become obese.[48] This was anecdotally demonstrated in a human, where an unfortunate woman who received a fecal transplant to treat her infectious diarrhea became massively obese afterward.[49]

Conversely, ingestion of certain strains of probiotics (friendly bacteria) or prebiotics (dietary components such as fiber that let friendly bacteria grow) have been associated with weight loss[50] and improvement in certain disease states.[51] Other studies have found probiotics can influence mood and cognition.[52] Microbial diversity may be protective against the various diseases of metabolic syndrome and obesity,[53] and possibly depression. Your bacteria talk loud and clear, and apparently your brain listens. Remember, 90 percent of the serotonin made in your body is used by the gut for various purposes; serotonin happens to be pretty versatile. Only 1 percent of your total body's serotonin is in your brain, impacting, among other things, your level of well-being and contentment. Apparently a happy gut means a happy you. These bacteria may alter our emotional state and our dietary preferences through indirect communication with the emotion centers within our brain.

The one undeniable fact is that of all the items in the grocery store, sugar is the only one that is independently associated with depression *and* addiction *and* metabolic syndrome. And, as we've already noted, it's undoubtedly the cheapest way to pleasure, and the surest path to unhappiness.

But fear not, there does appear to be one dietary item that can mitigate the damage that sugar does to the brain and promote the biochemistry and the processes that can predispose us to happiness. And perhaps not surprisingly its presence in the diet correlates positively with tryptophan and negatively with sugar. What is this magic chemical? It's omega-3 fatty acids, of all things. A type of fat. Something we were told to avoid forty years ago. Another item that's pretty hard to come by in the Western diet. Perhaps this is another reason that happiness has eluded so many of us.

Brain Food

Omega-3s come in two main flavors: eicosopentanoic acid (EPA) and docosohexaenoic acid (DHA). Everyone thinks these two omega-3s are found in fish. They are, but there are some qualifications to that statement. Fish don't make omega-3s; fish eat omega-3s. Rather, omega-3s are made by green leafy plants either in the sea or on land. Algae are the best source of omega-3s around. The fish eat the algae. We eat the fish. So we purchase our omega-3s secondhand, and at a premium to boot. But *wild* fish eat algae. Farmed fish eat pellets. Sometimes the pellets are made from other fish, which, even if they ate algae, are now pretty diluted. Sometimes they are made from corn. Since farmed fish are fatter—see, it happens to them as well!—they have a slightly higher omega-3 content, but their omega-6 content (which drives inflammation) is extremely high; thus wild fish is a more expensive but smarter choice.[54]

The pop literature surrounding omega-3s has translated into people popping fish oil capsules left and right. Full disclosure: I do too. There's even a purified omega-3 preparation available by prescription (Lovaza), but it's very expensive and usually not covered by insurance. In fact, I take care of children with severely elevated blood triglycerides—high enough to cause spontaneous acute pancreatitis, which is a mega disaster with life-threatening complications. Purified omega-3s are the obvious treatment of choice—and still the insurance companies won't authorize their purchase.

So what's behind this miracle superfood that does so much? Omega-3s are incorporated into cell membranes throughout your body. They increase "membrane fluidity," which means that they allow for easy cell deformation, allowing them to snap back instead of rupturing. This prevents cell aging and early cell death. They also allow nutrients and hormones to pass through the cell membrane, and allow toxins to leave the cell rapidly. Nowhere in the body is this special function more important than in the brain. For instance, omega-3s help repair the damage to the membranes exerted by glucose, and especially fructose.[55]

That's all well and good, but what do omega-3s have to do with serotonin, or with the promotion of happiness, for that matter? Turns out omega-3s impact our mental well-being in two distinct but related ways.[56]

1. Omega-3s have an indirect effect on serotonin release from nerve terminals throughout the brain. When the area surrounding the nerve terminal releasing serotonin is inflamed, it inhibits serotonin release (keys) and even fewer are able to make their way to the receptors (locks) across the synapse. This may explain why people whose bodies and brains are undergoing inflammation tend to be so irritable,[57] even if they are taking an SSRI. There is even less serotonin for them to work with. But omega-3s inhibit the formation of inflammatory cells,[58] which presumably would allow for better serotonin transmission.[59]

2. DHA, one of the omega-3s, is a precursor of a class of molecules called endocannabinoids (ECs)—the brain's and body's version of marijuana.[60] As we discussed with rimonabant in Chapter 2, we have specific receptors for marijuana, called CB1 receptors, which are ubiquitous throughout the brain. The active compound in marijuana known as tetrahydrocannabinol (THC) binds to this

CB1 receptor to heighten mood by alleviating anxiety, which explains why people are so giddy when they smoke pot. Turns out our neurons make their own marijuana-like neuromodulator called anandamide, which binds to that CB1 receptor, which is designed to alleviate our level of anxiety. Anything that inhibits anandamide synthesis or action will increase your level of anxiety severalfold, while anything that improves anandamide's action will keep you cool as a cucumber. Most of us are prone to experience anxiety and stress; an extra boost of occupancy of your CB1 receptors by smoking pot might allay some of that anxiety, if you aren't prone to paranoia when stoned. And, as you might expect, as our collective stress and anxiety levels have continued to increase, our level of happiness has continued to decrease. The increase in anxiety is one reason for the increasing number of potheads nationwide[61] and provides a rational explanation for why recreational marijuana legislation is spreading throughout the United States. Our own anandamide clearly isn't enough to tame the wild beast anymore. Why not?

Omega-3s are part of the endocannabinoid signaling machinery. Deficiency of omega-3s doesn't allow endocannabinoids to act as they normally would on the system, thus causing more anxiety and depression.[62] But the converse also seems to be true: we can fix this problem with omega-3 supplementation. In one study, a Mediterranean diet improved symptoms of depression.[63] Was it the omega-3s? Or less processed sugar-laden food? One study showed that omega-3s were equivalent in effect to Prozac in treating depression, and the combination was more effective than either one alone.[64] In a related study, administration of omega-3s to patients with recurrent self-harm (e.g., cutting, picking, scratching, burning—the ultimate expression of anxiety) showed a reduction in suicidality, depression, and daily stress.[65] A recent trial gave omega-3s along with minerals to eleven-year-old kids with conduct disorder or opposi-tional defiant disorder (the ones who routinely find themselves in the principal's office), and within three months their aggression was reduced, and way better than talk therapy.[66] Lastly, omega-3 consumption can help ward off depression in children[67] and adults,[68] and can serve as an adjunct to SSRIs in its treatment.[69]

And this is the gift that keeps on giving—or should I say the punish-ment that keeps on punishing? What *your mother ate* makes a difference in *who you are*. Many mothers state that their child's health and happiness are at the forefront of their own life goals. What you eat when pregnant plays a large part in determining your child's future. Lack of omega-3s during pregnancy in rats alters the offspring's brains in a way that messes with insulin signaling and brain growth factor levels, all of which leads to increased anxiety behaviors.[70] This has immediate implications for all of us. What do we tell pregnant women not to eat? Seafood—because of the concern for mercury poisoning. Except that maternal seafood con-sumption predicts improved neurodevelopmental outcomes in British children.[71] Are we making more trouble than we are solving? Well, maybe we can skirt this issue by giving the pregnant mom some omega-3 capsules; if we do, the kids' neurodevelopmental outcomes and the mom's risk for depression are improved.[72] New research suggests that walnuts might also be beneficial.[73]

We Are Such Stuff as Dreams Are Made On

There you have it—the dietary trifecta of happiness: tryptophan, sugar, omega-3s. Three individual components in our diet—two of which are essential nutrients and hard to get (tryptophan and omega-3s), and the third, added sugar, which is not even a food but nonetheless has been purposefully placed in virtually everything you eat and drink. Three separate mechanisms, but with a very clear interplay among all of them. The balance of these three molecules is the difference between a healthy, happy, and agile brain and an angry, sad, and demented one. And maybe your child's angry, sad, and cognitively compromised brain as well. The two that help alleviate depression are in short supply anyway—and even less so in processed food. Conversely, the component that destroys the brain has been added to virtually every processed food item for palatability and sales. Our Western diet has done us no favors in our ever-elusive quest for happiness. But just turn the page: the true key to our unhappiness lies in our insatiable quest for pleasure. It gets worse before it gets better.

10.

Self-Inflicted Misery: The Dopamine-Cortisol-Serotonin Connection

A lifetime of happiness! No man alive could bear it—it would be hell on earth." So says the character Tanner in George Bernard Shaw's *Man and Superman* (1903). A lifetime of contentment is reserved for the Dalai Lama. The rest of us mere mortals cycle through bouts of anxiety and dysphoria, squeezing out what pleasures we can and barely pausing long enough to enjoy any contentment we might be able to muster. But, as with everything else in this book, it's really about taming your biochemistry.

What if we could hype up our brain serotonin? What if we ate eggs and fish all day, consuming all the tryptophan and omega-3s in sight, and dumped the sugar? What if we managed to sleep a full seven hours a night? What if we fed our microbiome so it thanked us instead of paying us back? Wouldn't we be extremely happy? We know tryptophan is not enough: if you've survived even one Thanksgiving with unpleasant relatives, you know that bingeing on turkey is far from a guarantee of a blissful evening. Can you binge on contentment? Maybe our serotonin receptors would start to down-regulate, just like our dopamine receptors did (see Chapter 5), and we'd end up with less of a signal for contentment after all. If so, we'd all be doomed, striving for happiness but never quite achieving it. Is that our fate?

In contrast to dopamine, serotonin neurons have certain features that protect us from descending into the abyss—but they also prevent us from ascending toward nirvana (at least without chemical enhancements). In Chapter 3, we noted that dopamine down-regulates its postsynaptic do-pamine receptors. Get a hit, get a rush over and over, and the number of receptors goes down to protect the neuron, starting the vicious cycle of tolerance and dependence, crash-landing into addiction. If you bludgeon dopamine receptors, those neurons get beaten into submission and can eventually die.

An Emotional Thermostat?

In Chapter 7 we noted that serotonin-1a receptors in the brain are low in those diagnosed with major depressive disorder (MDD) from genetic studies;[1] from PET scans in live patients;[2] and biochemically in dead ones.[3] Yet we also noted that SSRIs fight depression by making serotonin clearance less effective in the synapse and increasing the odds of any sero-tonin molecule binding to a receptor. But if there's more serotonin at the synapse, shouldn't the -1a receptors down-regulate?[4] Why don't SSRIs stop working over time? Because, in contrast to dopamine receptors, the postsynaptic serotonin-1a receptors don't down-regulate in response to increased serotonin.[5] These neurons possess two special characteris-tics that keep our serotonin neurons and receptors resilient, even when we're not.

1. Serotonin neurons in the dorsal raphe nucleus (DRN) possess an extra control system: they express a set of serotonin-1a "autoreceptors" on the presynaptic side (the neuron releasing the serotonin). What does this mean? These receptors normally serve as a feedback loop that regulates how frequently that neuron fires. It's like the servo-mechanism of the thermostat in your house. When the temperature drops, the thermostat kicks the heat on, and when it gets too hot, the furnace turns off. The -1a autoreceptor serves as the neuron's thermostat, causing it to fire relatively slowly and rhythmically, and silencing it before it gets into trouble. By preventing these neurons from firing too rapidly, these autoreceptors make sure the serotonin neurons don't wear out; and there is rarely enough serotonin in the synapse to down-regulate those -1a autoreceptors. SSRIs turn those autoreceptors off; it's like setting the temperature threshold on your servo-mechanism thermostat much higher, and those serotonin nerve terminals fire like gangbusters; thus the antidepressant effects.[6]

2. Perhaps the most amazing thing about serotonin's binding to the postsynaptic -1a receptor is that, rather than stimulation, serotonin *inhibits* the next neuron.[7] Postsynaptic -1a agonists quiet the postsynaptic neuron, giving them a rest. Remember from Chapter 5 that neuronal death occurs from a process called *excitotoxicity*, when a neuron keeps firing and kills its target. But there is no such thing as *inhibitotoxicity*!

What's more, the serotonin system has one more trick up its sleeve: it has the capacity to tame (or excite) the dopamine system.[8] Of course, we need both. Personally, the idea of a life with contentment but without motivation and reward, living atop a mountain in deep meditation and pondering how much karma I have lost driving in New York for twenty-five years, just isn't all that appealing. But neither is dopamine overload and its aftereffects. How to get a balance? Serotonin agonists such as LSD and psilocybin (magic mushrooms) are being used as potential treatments for smoking, alcohol, and other drug addictions. Family and religion can serve the same purpose. Could happiness and contentment reverse addictive behavior? Could our own serotonin overcome our dopamine to our benefit? Indeed, the animal data say yes, serotonin can speed up the breakdown and disposal of dopamine,[9] resulting in decreased dopamine-related reward signaling and reward-seeking behavior.[10,11,12] In humans, we know that high levels of serotonin can decrease alcohol intake[13] (although your antidepressants can't work if you're imbibing three bottles a day). Conversely, serotonin depletion (through experimental depletion of tryptophan) is associated with risky choice making.[14] Could the food deserts and overall poor nutrition in inner cities impact the crime rates? Story for another time, but they certainly don't help.

Your environment, like your genetics, can make a huge difference in the functioning of this system. Low dietary tryptophan means less serotonin gets made. Fewer -la receptors means serotonin can't do its job. Fast serotonin recyclers/transporters (the hungry hungry hippos) mean less of a chance for each molecule to get to the receptor in the first place. Any of these three things can lead to depression (see Chapter 7). And what if these serotonin neurons die? The serotonin system is not impervious to damage; it's just that (unlike dopamine neurons) serotonin isn't the likely culprit.

"Breaking" Neurons Is "Bad"

Serotonin does not exist in a vacuum. It is both directly and indirectly impacted by different neurochemicals, including other drugs, cortisol (stress), lack of sleep, and crappy diet. All the things that negatively affect dopamine as well. Uh-oh, can't you see what's coming?

Many illicit party drugs tap-dance on both your serotonin and dopamine systems. While onetime cocaine exposure might provide a serotonin boost, binge cocaine administration does anything but. The chronic blockade of the DAT (dopamine's hungry hippos) also plays havoc with your dopamine receptors (due to tolerance), but it's also knocking down the serotonin-1a receptors in key regions that matter,[15] which means that the happy rush is now a pretty sad puddle. Al Pacino's Tony Montana in *Scarface* (1983) was anything but Zen at the end.

But binge cocaine use is milquetoast compared to industrial-strength methylenedioxymethamphetamine (MDMA), the recreational drug known worldwide as "Molly" or "ecstasy." This synthetic neurotransmitter analog has been available since the 1980s and has slowly inched up the list of substances of abuse that are creating societal problems. MDMA is the ultimate club drug, because it provides the user a panoply of neural experiences all at once; it is the ultimate reuptake inhibitor. It binds up the hungry hippos of both dopamine and serotonin and puts them both out of commission. In other words, dopamine and serotonin run full tilt at the same time. It heightens excitement and sexuality and postpones fatigue and sleepiness, because the dopamine receptor is activated; it increases euphoria, because the serotonin-1a receptor is activated; and it even gives the added bonus of minor hallucinations, because the serotonin-2a receptor is activated,[16] although the bonus "mystical experience" is not part of the portfolio.[17] Three big bangs for one buck, as it were. Yes, MDMA has it all—sex, drugs, and rock 'n' roll. Except for one additional bonus: long-term MDMA use kills neurons. And not just the postsynaptic cortical neurons, which are responsible for the defects in memory, decision making, and impulse control. No, MDMA can kill the DRN serotonin neurons outright, and scar the brain, by enacting the same program of cell death that cocaine does.[18, 19] And the current drug of the day—methamphetamine—will also kill off the nerve terminals of both dopamine and serotonin neurons.[20]

Yet, like LSD, whose early indiscriminate use ultimately paved the way to more controlled research and potential benefit, MDMA might be useful, and researchers are beginning to examine those scenarios. We know that people with autism, social anxiety disorder, and post-traumatic stress disorder are mentally and socially compromised. Life for these patients is extremely difficult to navigate, in part because they have trouble connecting emotionally with others. But early controlled studies administering MDMA as single doses to autistic adults in medical settings are demonstrating increased openness, introspection, and social adaptability,[21] without starting the slide to addiction. The improvement in mood and reduction in defensiveness is allowing treated individuals to join social structures and participate in rewarding behaviors, such as dancing. The effects, like single-dosage LSD, seem to be long-lasting, and with few if any side effects. Presumably, the boost in dopamine reduces the fear of social anxiety, while the boost in serotonin increases contentment, providing the impetus to let these people participate in society.

I hear you already: like *South Park*'s Mr. Mackey, you're saying, "Drugs are bad, mmmkay?" Indeed they are, when taken indiscriminately and over the long term. Everyone's reaction is different; contrary to popular belief, your brain is actually the most sensitive part of the body. But there are way more people who wish they were happy with a nonexistent magical pill, because unhappiness (exclusive of clinical depression) hovers at 43 percent of all Americans, at least those who admit it. They're on the same spectrum as those who are clinically depressed, they're just not as severe. What are their serotonin systems up to? In one study, the MRI findings of people who identified as being less happy had on average fewer serotonin transporters or serotonin-1a receptors.[22]

Stress Pushes Us over the Edge

OK, clinical depression is roughly 7 percent of the population, drug addiction requiring rehab is about 9 percent . . . but everyone is stressed nowadays, and for most of them, it's chronic stress. In Chapter 4, we learned that stress and dopamine feed each other in a "positive feedback" cycle. Remember the prefrontal cortex (PFC), the "Jiminy Cricket" part of the brain that's supposed to inhibit impulsive behavior by turning off the amygdala? Dopamine nerve terminals reside in the PFC and are ideally kept in check. But a massive amount of stress-induced dopamine flooding the PFC will squash that Cricket, increase risky and impulsive behavior,[23] and keep your cortisol elevated.

As you might expect, cortisol is the anti-contentment hormone. Contentment means all is well and it's OK to chill. If the adrenal glands are releasing cortisol, something must be wrong: time to get some butt in gear—mobilize the glucose, mobilize the fat, grab a chain saw and prepare yourself for the zombie apocalypse. OK, too much *Walking Dead* (2010 to the present), but you get the idea. This ain't chillin' time.

Female monkeys, like humans, form social hierarchies. Some are pretty impervious to stress, so no big surprise that they are at the top of the social totem pole. Those who are more subordinate, who have to scramble for food and status, have the equivalent of a more stressful life. And their serotonin gets shot down too. They have fewer serotonin-1a receptors in the DRN.[24] Indeed, stress and cortisol are the mortal enemies of the serotonin-1a receptor, and cause down-regulation throughout most species—in the Gulf toadfish,[25] in the rat,[26] in the tree shrew,[27] and in the human.[28] Fewer -1a receptors means less serotonin signaling, and less contentment.

Depressed people have problems with circadian (day-night) cortisol regulation. Normally, cortisol goes up in the morning before you wake up to help you mobilize glucose, raise your blood pressure, and get ready for the day. By the time nighttime arrives, cortisol levels are in the sewer. This circadian rhythm of cortisol is missing in depressed subjects: their cortisol is *always up* and you can't even suppress it with medications, making this a tough nut to crack. One study suggests that cortisol reactivity may be a predicting factor for suicide.[29] More stress means more cortisol, which plays havoc by down-regulating the -1a receptor, reducing serotonin signaling and increasing risk for depression and apparently for suicide as well.

And those who are cursed with a specific genetic difference making lots of serotonin recycler/transporters (hungry hungry hippos) are at the highest risk. If your hippos are gobbling most of your serotonin in the synapse, and stress is eradicating your -1a receptor, you are totally screwed.[30] The worst of all chronic cortisol problems stems from adverse childhood experiences, or ACEs (better known as child trauma).[31] Pick your abuse (physical, sexual) or your stress (parents' divorce, fighting, bullying): ACEs can result in cortisol dysregulation into adulthood,[32] including increased risk for addiction and depression. Kids who experienced ACEs and who also harbored a genetic difference in their serotonin recycler/transporter demonstrated a fourfold risk in depression when they hit adulthood.[33]

And what does America suffer from across the board? Chronic sleep deprivation. If you really want to make someone unhappy, deprive him or her of sleep. Given that roughly 35 percent of the adult population doesn't get enough sleep (on average less than seven hours per night),[34] what effect does this have on happiness? When I was in med school, I frequently pulled all-nighters, and I was anything but pleasant. Screen time, stress, work/life balance—it's harder than ever to actually fall and stay asleep. Arianna Huffington has argued for a sleep revolution ("Ladies, we are going to literally sleep our way to the top!")—possibly the least violent revolution in the history of humanity, but arguably one of the most important.

Sleeping Your Way to the Bottom

What is the relationship between sleep, eating, irritability, and serotonin? Well, like everything in the brain, it's complicated. Chronic sleep deprivation is a hallmark of severe aggression and irritability in some people[35] and suicidal depression in others.[36] Those who suffer from depression generally have off-kilter sleep habits, either sleeping too much (hypersomnia) or sleeping too little (insomnia). Brief shout-out to all those new parents who managed to survive those first two years.

Chronic insomnia is a major risk factor for depression.[37] Whether chronic sleep deprivation affects serotonin receptors directly or indirectly through effects on cortisol is not yet known, but mouse studies would suggest that cortisol plays some role.[38] In rats, lack of sleep messes with cortisol reactivity to stressful situations[39] and simultaneously decreases the function of serotonin-1a receptors.[40] As you might suspect, human data are hard to come by, as most institutional review boards frown on Gothic research methods such as long-term sleep deprivation.

Not only does insomnia and sleep deprivation wreak havoc on our mood and emotions, it messes with our waistlines, increasing insulin resistance, glucose intolerance, obesity, and all the other diseases of metabolic syndrome.[41] One study found that sleep deprivation increased wanting of high-caloric food items (compared to when the participants had a good night's sleep), and these changes in behavior corresponded with decreased activity in brain regions related to decision making but increased activation in the amygdala (the stress and fear center).[42] And it likely goes both ways. Adults who get five or fewer hours of sleep per night consume 21 percent more sugared beverages (including energy drinks) than the general population, while those who sleep six hours or less consume 11 percent more.[43] But, as in all other correlation studies, what drives what? Do sugar and caffeine cause sleep deprivation? Or does sleep deprivation drive sugar and caffeine consumption? Either way, when you're chronically underslept, the Cricket gets stomped, and it's time for Taco Bell's "Fourth Meal."

The Most Unhappy of Pleasures

Indeed, one of the most pernicious causes of unhappiness across populations is bad food. Remember from Chapter 9 that tryptophan is low in
the fast-food diet, yet the competing amino acids tyrosine and phenylalanine are abundant. More precursors for dopamine means more occupied
transporters, which means less chance for tryptophan to get across into
the brain to be converted to serotonin. It would seem that any pleasurable item we consume that drives dopamine up (sugar, alcohol, processed foods) can also drive serotonin down, possibly directly, or indirectly
through metabolic syndrome. Conversely, weight loss that reverses metabolic syndrome can also improve symptoms of anxiety while at the same
time increasing blood serotonin levels,[44] although blood levels are not necessarily relevant.

Depression rates started to increase after the post-1940 birth cohort
reached adulthood—in the 1960s,[45] at the same time processed food
started its ascendency as the world's diet staple. While this temporal relation is not causation, it's still pretty suspect.

In this chapter, we have identified all the components leading to
unhappiness and how they interact to ruin our mental well-being. Serotonin keeps dopamine in check, yet it appears that the same things that
raise your dopamine can also tank your serotonin. Add cortisol to the
mix, and happiness becomes unattainable. Figure 10-1 demonstrates how
our current environment can contribute to our anguish. And it can happen to anyone. The worst part is that the triggers to enter this pathway
are all around us: reward and stress are the hallmarks of modern civilization. This is our seesaw pathway to misery, the fulcrum on which our
entire society rests. And once you enter, it's difficult to escape. Doctors
prescribe SSRIs while people self-prescribe marijuana, all to accomplish

the same effect; no wonder they're both such hot sellers. But there are other, more sustainable solutions (see Part V).

Nature wired our brains for both contentment and motivation, two seemingly dichotomous states. Everyone wants to believe they have free will—that they have choice over their own actions. Then why would anyone freely choose addiction or depression, with more "choosing" them every day? Addiction and depression are not choices that people make willingly. Our environment has been engineered to make sure our choices are anything but free. It chronically nudges us toward reward and drives us away from happiness and contentment. These ostensible choices have obvious personal costs, but they also have societal costs as well. Part IV will elaborate on how all of this science impacts our seemingly conscious choices, how government and business have used this science against us to manipulate these choices, and how our choices negatively impact society at large.

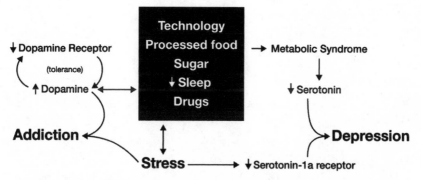

Fig. 10-1: The road to hell is paved with good intentions. The same factors that increase dopamine (technology, lack of sleep, drugs, and bad diet) also decrease serotonin. Furthermore, stress drives dopamine release and also decreases the serotonin-1a receptor reducing serotonin signaling. Addiction results from dopamine receptor down-regulation coupled with excessive stress. Depression results from reduced serotonin transmission from the same precipitating factors, also coupled with excess stress.

PART IV

Slaves to the Machine: How Did We Get Hacked?

11.

Life, Liberty, and the Pursuit of Happiness?

"N o man is an island," wrote British poet John Donne (1572–1631). Each of us influences everyone else around us, whether in our family or our community. Wise words. The quote continues: "Any man's death diminishes me, because I am involved in mankind; and therefore never send to know for whom the bell tolls; it tolls for thee." If Donne is right (and he is), our personal happiness is tied up in the general level of contentment or distress within the rest of the population. If I am unhappy, is everyone around me unhappy? And if they are unhappy, was I the cause, or is there an outside force detrimental to each and every one of us?

Chapters 3 to 10 delineated the reward and contentment pathways as: (1) distinct, (2) overlapping, (3) regulatable, and most importantly (4) interactive. We're told that things will make us happy, but they don't. We're told we should be ecstatic, but we're not. Because what we're told is based on a faulty premise—that pleasure and happiness are one and the same. A

premise ingrained in the American psyche, and indeed throughout Western civilization. Industry and government call it economic progress, but it
is they who have subverted the meanings of these two emotions—reward
versus contentment, pleasure versus happiness—for their own purposes.

And we bought the subversion, both figuratively and literally. It's the
bedrock on which our economy is built. We spend money on hedonic
pleasures, trying to make ourselves happy, and in the process we drive
dopamine, reduce dopamine receptors, increase cortisol, and reduce serotonin, to ever further distance ourselves from our goal. The cognitive
dissonance between our expectations and our reality is deafening. But
what happens when this cognitive dissonance is societal, on a grand
scale? What happens when you go from "It's all about *me*" to "It's all
about *we*"? The science that belies the interaction between reward and
contentment fuels both individual and societal unhappiness.

In Chapters 11 to 15, I will demonstrate, in turn, how the confusion
between these two terms in the name of "progress" has inflicted personal,
economic, historic, cultural, and health/health care detriments to individuals and to society in general. Moreover, this confusion continues to be stoked
by industry and government in order to preserve and sustain persistent economic growth at the expense of the populace. Some would argue that a
strong stock market and an increasing GDP must mean that we're on the
right track. Yet, even before the 2016 election, three-quarters of the country
thought we were on the wrong track.[1] This is flagrant cognitive dissonance.
Why? We can't possibly fix the problem if we don't understand it.

The Abdication of the Declaration

Thus far we've equated individual happiness with *eudemonia*, or contentment. What does societal happiness look like? Is collective unhappiness
also driven by dopamine, cortisol, and serotonin? How do you define it?

How would you measure it? How do you know if you've run off the rails? And if you have, how would you fix it?

To demonstrate this cognitive dissonance on a societal level, let's take a familiar example: America. We have the best system of government ever devised, although lately it has been put to the test. We have more natural resources than any other country. We have personal liberties that are unparalleled; we have a constitution that in the past has equalized personal versus governmental authority that has allowed the individual to flourish; and we have a series of checks and balances that, while often acutely fallible, over time have tended to right themselves. Winston Churchill said, "You can always count on America to do the right thing—after they've tried everything else." Witness the zigzag successes of the civil rights struggle, changes in gender and now transgender equality, and same-sex marriage. Although most of the world still envies us, we've recently taken a hit due to the U.S. government's foray into isolationism and protectionism. One reason America remains a prime target for terrorism is that the internet and social media now demonstrate to the "have-nots" what it looks like to be the "haves"—and they've subscribed to the premise "If you can't join 'em, beat 'em." Yet for all of its benefits, fairness, and provisions for economic opportunity, America is very unhappy.

Our Declaration of Independence assures us of the rights to "life, liberty, and the pursuit of happiness." Well, "life" is going in the wrong direction, as witnessed by an increase in the American death rate[2] and a decline in the mean American life span.[3] Compared to other developed countries, American life expectancy is not all that terrific—only twenty-sixth out of the thirty-seven in the Organization of Economic Co-operation and Development (OECD: the most wealthy countries). But now U.S. federal data drawn from all deaths recorded in the country show that life expectancy for women fell from 81.3 in 2014 to 81.2 in 2015; for men, life expectancy fell from 76.5 to 76.3 years. And this is true across demographics. Age-adjusted death rates increased in 2015 from 2014 for non-Hispanic black males (0.9 percent), non-Hispanic white males (1.0 percent), and non-Hispanic white females (1.6 percent).[4] This is a measure that economists use to quantify societal health, and it is the first time in recorded history where U.S. life span has declined.[5] While it doesn't seem like much, statisticians argue that this is a watershed moment in our country's history.[6] In addition, between 2014 and 2015, America saw an increase in infant mortality.[7,8] Despite our prowess in advanced health care technology, America ranks thirty-fourth in perinatal mortality among all countries. Black babies are twice as likely to die at birth than white babies. These indices are used by policy makers to gauge the vibrancy, health, and stability of individual countries. Their deterioration foreshadows a very uncertain future for America.

"Liberty" is a mixed bag. We have many social freedoms, yet we are stuck in metaphorical jails of our own construction, whether they are private gated communities or urban ghettos. Even when we try to pry ourselves out of them, we haven't really escaped. Harvard economist Raj Chetty has shown that children born in poor neighborhoods have only a 1 in 10 chance of being able to improve their financial situation and social standing. The neighborhood you grew up in predicts your chance for upward mobility.[9] For example, each additional year that a child spends growing up in suburban Chicago, Illinois, raises the household income in adulthood by 0.76 percent. That's worth a 15 percent annual salary bonus at adulthood compared to the national average. Conversely, each additional year spent growing up in inner-city Baltimore lowers the annual earnings by 0.86 percent. For a Baltimore native, that's worth a 17 percent annual reduction in salary by adulthood. In the immortal words of the Eagles frontmen Don Henley and the late Glenn Frey, "You can check out any time you like, but you can never leave."

And "happiness"? The hard data of the increase in death rate and decline in life span serve as markers of our societal unhappiness. With the obesity and metabolic syndrome epidemics in full force and continuing to worsen, common sense would dictate that these excess deaths would be occurring at the upper end of the age pyramid, and that type 2 diabetes and all of its complications would be knocking off the elderly. But that is not what is seen. Rather, these alarming mortality statistics are being driven by the deaths of white Americans. While America has been preoccupied with the continuing specter of the murders of young black men and police officers, deaths from drug overdose have reached an all-time high.[10] In 2015 there were more deaths from opioid overdose than there were from shootings.[11] Nobel prizewinner Angus Deaton and economist wife Anne Case have tracked this demographic group relative to other countries and relative to other racial groups. This increase in deaths was largely ascribed to: (1) acute self-inflicted poisonings, which suggest suicides, (2) accidental overdoses of prescription and street opiates in people who had become addicted, and (3) cirrhosis of the liver due to chronic alcohol abuse.[12] And all of these preventable deaths are being tallied in white America,[13] the demographic with the highest income of any group on the planet (although the data argue that within these white American statistics the highest mortality rates are in those with only a high school education). Deaton and Case point to a "cumulative disadvantage over life," which is unrelated to income, but exacts itself through the labor market, in marriage and child outcomes, and in health decrements.[14] Since the advent of Prozac in 1987, suicides and suicide attempts have declined slightly in the U.S.[15] as a whole; yet, in the otherwise well-to-do white middle-class demographic, the opposite is apparent, with a 2.3 percent increase in the suicide rate in 2015.[16] If we're pursuing happiness, we sure aren't catching up to it.

Thomas Jefferson vs. George Mason vs. the Pursuit of Property

How did we get to this place in time? How and when did happiness become subverted by pleasure? A brief but deep dive into American history will help, because you can't understand the message unless and until you understand the messenger(s). Our politicians routinely cite the Declaration of Independence and the Constitution, interpreting the words and the intentions of the founding fathers. But who were these men and why did they craft these specific cornerstones of our democracy? That last clause of the Declaration, "the pursuit of happiness," has a very checkered history. It was mentioned twice in print in 1776, and then it disappeared, never to be written again. In its place, the U.S. Constitution replaced the word "happiness" with "property." The acquisition of the tangible quickly superseded the quest for inner contentment.

What happened to happiness? How did it get stricken from the record? The pursuit of happiness is an inconsequential anachronism today: conceived of our best intent, yet reduced to our worst ignominy. The famed economist and pundit Robert J. Samuelson in 2012 wrote, "The happiness movement is at best utopian; at worst, it's silly and oppressive."[17] He argued that the pursuit of happiness may be a right guaranteed by the Declaration, but its achievement is not an entitlement. Maybe he's right, because we were only guaranteed its "pursuit." Instead, the due process clause of the Fifth Amendment of the Constitution assures us of the government's inability to deprive us of our right to "life, liberty, and property." Property? Is property the booby prize for happiness? Or is property (as stated in the Fifth Amendment) just the guarantee of an individual's right to acquire those necessities in life that can ensure happiness? In other words, is property a means to an end, or an end in itself? Is property necessary to one's happiness? Can one be happy without property, or does one need it to be happy at all? Is happiness rooted in material possessions?

Thomas Jefferson, the primary author of the Declaration of Independence, was a master of eloquence and brevity. He charged our thirteen separate colonies to organize for a common cause and a set of common goals. He imbued our nascent country, on the verge of a revolution never accomplished in world history, to soaring aspirations—to be better than we were. Happiness was right up there with life and liberty. Not a word about pleasure. When you're staging a rebellion, the central tenet can't be about dopamine—nothing so sordid as a pint of rum (the substance of abuse of the time), which apparently was the coin of the realm back in 1776.

Yet it turns out that Jefferson lifted the line about happiness from his contemporary George Mason, who composed the Virginia Declaration of Rights, which was adopted by the Virginia legislature just three weeks before the fateful vote that established the Union. Mason was a student of history but not of creative writing or oratory. He drew on a long tradition of statements of rights and liberty including the Magna Carta (1215), the Habeas Corpus Act (1679), and British statesmen John Locke's *Second Treatise on Civil Government* (1690). Mason's actual verbiage in the Virginia document reads: "life and liberty, with the means of acquiring and possessing property, and pursuing and obtaining happiness and safety." None of Mason's primary sources mentioned happiness; he threw that in on his own. And Mason didn't just want to be able to pursue it but to actually obtain it (at least for some people). A unique mind-set for those times, to be sure.

Jefferson liked the idea of including happiness, borrowed Mason's line, and reworked it for his Declaration, but then he deleted the property clause. Why would he do that? Didn't Jefferson believe in property? Or did he consider pleasure and happiness to be the same thing? More has been written about the moral vicissitudes of the enigmatic Jefferson than virtually any of our founding fathers (although Alexander Hamilton got the musical). Jefferson had more layers than a Shakespearean tragedy merged with a John Grisham novel. He was a patrician with a big bankroll, spent years living in luxury in France, and, most importantly, was a slave owner. And back in 1776, slaves were property. You'd think he of all people would have defended the right to accumulate and maintain property. But Jefferson deep-sixed the concept of property in the Declaration. Was deleting the property clause from Mason's original text Jefferson's method of absolving himself of his own shame and moral guilt that he felt as a slave owner? Making peace with his slave ownership? Was he perhaps advancing the cause of abolition? Or did it simply imply that you couldn't very well pursue happiness unless you had property, including slaves? Many books have been written on this subject, perhaps the most revealing by historian Henry Wiencek.[18] It's not clear whether Jefferson truly abdicated happiness for property, because he couldn't rescue the clause when the Constitution was drafted, since he was serving as ambassador to France.

Mason was also a very curious character in the history of our Union. He was the first to mention happiness as a goal of life and of government; yet happiness eluded him most of his life. His father drowned when he was ten, and his mother brought him up alone. He was a very bright student, and although he wasn't formally educated at the bastions of learning of the day, he had resounding success in business dealings throughout his life and he became an extremely rich man, spending quite a bit of energy amassing property. His estate, Gunston Hall, was right next door and second in size to Mount Vernon, and he and George Washington had several business dealings before and throughout the Revolutionary War. He was married with ten children; however, his first wife, Ann, passed away giving birth to twins in 1773. In between, he was a Virginia statesman when he wrote his Virginia Declaration of Rights, in which happiness was a primary goal. No doubt his loneliness and possible depression during this interval weighed heavily on his mind, until Mason married his second wife, Sarah Brent, in 1780.

At this point, a perusal of Mason's medical file is in order. Mason was quite affluent, but apparently his money couldn't buy him happiness. In fact, he was in pain for most of the second half of his life. Check out the paintings of Mason from his legislative days on Google Images. He was quite obese and, more importantly, he suffered from gout. The bronze statue of Mason at George Mason University in Virginia has him leaning on his desk, favoring his left foot, due to the pain in his ankle from the disease.[19] Gout is a chronic inflammatory disease of small joints due to the deposition of the liver waste product uric acid (see *Fat Chance*). Obesity and gout are both co-morbidities of metabolic syndrome, the cluster of diseases that signify metabolic dysfunction, such as type 2 diabetes and heart disease (see Chapter 10). What are the two causes of gout? Sugar and alcohol.[20] In the 1770s these were two of the three readily available substances of abuse (tobacco being the third). But Mason was a teetotaler, so his gout and obesity weren't due to alcohol. Mason knew that sugar was a primary cause of gout, because his contemporary Benjamin Franklin, a fellow sufferer from gout and other symptoms related to metabolic syndrome, knew that his sugar and rum consumption were the cause; he even wrote a famous poem about it.[21] Similarly, Mason suffered from poor health for the twelve years after marrying Sarah in 1780. He refused to travel out of state, and in 1789 he declined a request by the state legislature to be the first U.S. senator from Virginia, as he would have had to travel back and forth to Philadelphia. Obesity and gout, both markers of metabolic syndrome and both consequences of his sugar habit, rendered him functionally disabled just as it does 25 percent of Americans today.[22] Not a happy guy.

Two statesmen who birthed a nation. Both Jefferson and Mason yearned to steer the country toward the pursuit of happiness as an establishing principle of the Union. Jefferson saw it as a primary goal, while Mason saw it as a secondary outcome that emanated from ownership. Yet both abdicated in the end. The pursuit of property and pleasure trumped the pursuit of happiness in 1788. Dopamine trumped serotonin. And it still does. No wonder we're unhappy.

"Good Night Sweet Prince, Flights of Angels Sing Thee to Thy Rest"

Aristotle argued "the pursuit of happiness and the avoidance of pain is a first principle; for it is for the sake of this that we do all that we do." Yet it looks like our avoidance of pain has taken precedence over our pursuit of happiness, and it's come home to roost in the guise of our national opiate crisis. Opiates mollify the perception of pain, but in the process they mollify the perception of life, and in the extreme they mollify life itself. There have been drug addicts for as long as there have been drugs. In the past, most victims tended to be young, misfit, and poor. Musicians of the 1960s like Janis Joplin and Jimi Hendrix were easy to dismiss as part of the prevailing counterculture. Lately such well-established talents such as Prince, Philip Seymour Hoffman, Amy Winehouse, and Heath Ledger are just a sampling of those who have recently succumbed. But this isn't happening only to entertainers: they just get the publicity. Deaths from drug abuse are increasing in just about every demographic—old, young, white, black, Latino, and now the middle-aged as well.[23] Just look at the unfortunate consequence of secondhand drug availability: inadvertent opiate poisoning in toddlers, which has more than doubled in the past sixteen years.[24] If toddlers are dying of accidental overdose, it's not just because famous and disaffected youth are the abusers.

Athletes have long been using steroids for their performance-enhancing effects, but some have graduated up to opiates, starting usage due to an injury and then continuing it for its mind-altering effects.[25] The same is happening with weekend warriors who tear an ACL or a muscle and end up using for the rest of their lives. A full 80 percent of heroin users started by using prescription painkillers first, and one out of fifteen people exposed to those painkillers will try heroin within the next ten years.[26] Listen, when *Saturday Night Live* does a parody on "Heroin AM,"[27] you know this phenomenon is mainstream. Why? Where did this come from, and why so fast? We recognize and bemoan the celebrity victims only too well, but do you recognize your neighbor? Your bus driver? Your kid? Your pilot? Who are the perpetrators of this national scourge and what's driving it?

Pain Doctor or Drug Pusher?

This opioid explosion among the white middle class has escalated over the past fifteen years, and surprisingly the fuse was lit by the least likely of all pushers: the medical profession itself. Time for another brief history lesson that you won't learn in high school—or in medical school, for that matter. For the first three decades after the Drug Enforcement Act of 1970, we physicians were counseled on how *not* to prescribe Schedule II drugs (i.e., those with addictive potential). People in genuine pain (e.g., cancer patients) had nowhere to turn for relief, even at the end of life. Fast-forward to 2001: A Drug Odyssey. In response to patient satisfaction questionnaires on physician behaviors as well as inpatient hospital feedback, it was clear that doctors were missing and/or not treating their patients' pain. The American Medical Association and the entire medical profession did a complete 180-degree about-face. Treatment of pain was only cursorily covered in medical school, and the only course of therapy was opiates. But all of a sudden, pain became a painfully common topic among physicians. Was this a natural evolution in physician attitude? Or something more sinister? The timing of this pain revolution temporally coincided with two external changes in the legal drug culture.

1. Prior to the year 2000, the strongest narcotics, such as fentanyl, morphine, and heroin, were only available intravenously, which left many people out of the drug culture—because train tracks were

grounds for losing your job or a Child Protective Services referral. However, in the late 1990s several potent oral opiates became available, such as OxyContin, Lortab, and Vicodin. Their introduction changed the playing field in several ways. The American Pain Society published new guidelines that advocated for doctors to aggressively treat pain. The Joint Commission on Accreditation of Hospitals, which certifies health facilities, issued standards in 2001 that instructed hospitals to measure pain: it became the fifth vital sign after heart rate, respiration, temperature, and blood pressure. But most importantly, the companies that manufactured these narcotics (Purdue Pharma, Johnson & Johnson, Endo Pharmaceuticals) began to aggressively market these opiates for muscle and joint pains, even long-term, on the premise that because they were long-lasting (like methadone) they were not addictive. They promoted prescription narcotics use through medical journals and continuing medical education courses, and even funded a report by the U.S. Government Accountability Office to give it the imprimatur of government complicity. State medical boards started to punish doctors for inadequately diagnosing and treating pain; and these same boards willingly accepted money from pharmaceutical companies to disseminate these new guidelines. Recently, unsavory narcotic marketing practices by Big Pharma have come under fire. One firm pled guilty to criminal charges that they had misled the FDA, doctors, and patients about the risks of narcotic addiction and abuse by marketing their drug as a safe alternative to standard narcotics.[28]

2. A new oral medication developed to combat opiate addiction called Suboxone (buprenorphine plus naloxone) was approved in 2002, giving physicians an added sense of security that any addicts could be easily be managed by preventing drug withdrawal. The advertisements from the pharma companies said that this combo drug reversed addiction because of the naloxone (which is an EOP

receptor blocker). But in fact Suboxone is addictive in naïve people because of the buprenorphine. Suboxone, the medical anti-opiate, became a hit on the black market as a legal opiate.[29]

The Centers for Disease Control (CDC) reports that deaths from opioid overdoses in the U.S. have quadrupled since 1999, with twenty-eight thousand deaths in 2014 alone, of which half were attributed to prescription pain relievers obtained legally, some of which were then sold as black market drugs. Pain clinics are now almost as common today as emergency "Doc-in-the-box" franchises in some cities. One-third of all long-term users say they were originally prescribed their opiates due to an injury, but that now they're addicted; and six out of ten say their doctor never told them how to stop.[30] All of this has occurred while 99 percent of our citizens' pocketbooks are growing to their highest level ever, even after controlling for inflation,[31] although certainly not after controlling for the incomes of the top 1 percent. If the backbone of the middle class finds itself squeezed economically, drug addicted, and dying young, it sure doesn't sound like America is particularly happy. The new iPhone just isn't cutting it.

The Whole Country's Going to Pot

Perhaps the most obvious display of America's current unhappiness is the state-by-state switch from one anti-anxiety agent to another. Five states and the District of Columbia have now legalized marijuana for recreational use, and another twenty-eight states approve its medicinal use. In Washington, Oregon, and Colorado, the three states for which we currently have data, the drop in SSRI use has been inversely proportional to the increase in pot use.[32] Short-term, marijuana has long been used to reduce acute anxiety, and many with terminal diseases have turned to pot to make the ride more bearable. But the question that still plagues us is:

What does long-term use do? Numerous studies document that frequent cannabis users have a high prevalence of anxiety disorders, and patients with anxiety disorders have high rates of cannabis use.[33] But again, is this cause or effect? We really need to know if marijuana use increases the risk of developing long-lasting anxiety disorders, yet sadly, the data continue to confound.[34,35] One lingering concern is whether long-term marijuana usage causes brain damage. While it would appear that THC stimulation of the CB1 receptor does not cause direct cell death, there is a burgeoning literature that marijuana use can alter brain networking and connectivity, especially in adolescents,[36] manifesting as altered educational or social development into adulthood.[37]

Even though marijuana is not covered by insurance and it's regulated and limited to purchases of one-quarter ounce at a time, costing about $150—it's taxed to the hilt—in these states it's legal without a doctor's prescription. The fact that thousands of people have turned their backs on SSRIs in favor of pot in order to deal with their anxiety says two things: (1) they're unhappy, and some may even suffer from undiagnosed subclinical depression, and (2) they're willing to spend a lot of money in an attempt to get happy. Furthermore, with the repeal of Obamacare, and people not being able to afford their medications, expect to see an uptick in usage. If there's good news here, it's that more widespread legalization means more competition and prices will fall. But then more and more cannabis farmers will cry foul, trying to maintain control over this cash crop. It ain't called the "green rush" for nothin'. When the U.S. government starts subsidizing marijuana (mark my words, it's coming), you'll know we've hit rock bottom.

12.

Gross National Unhappiness

When were we happy? The icons of the 1950s, such as *Leave It to Beaver* and *Ozzie and Harriet* and *Father Knows Best* and *I Love Lucy*, suggested that husbands going to work, wives keeping house, kids with the usual adolescent angst and growing pains— i.e., family life—ostensibly was the Golden Age of American Happiness. Except the TV version wasn't real. Actor Robert Young suffered from depression and alcoholism, and Desi Arnaz was a serial philanderer. Former Harvard president Derek Bok states that the percentage of Americans describing themselves as either "very happy" or "pretty happy" peaked in the 1950s, and thereafter has remained virtually constant, irrespective of rises in absolute or relative income.[1] Were we really happier in the 1950s? Or is it that we were just told that we were? What happened to alter our perception of happiness? Or were we ever really happy? *USA Today* documents that our view of our happiness has not changed appreciably between 1972 and 2010: those responding "very happy" went from 30 to 29 percent, "somewhat happy" from 53 to 57 percent, and "not too happy" from 17 to 14 percent. Back in the 1950s we had minorities not voting, women not competing, and people not knowing what their government was doing. Even Adam and Eve were happy until they got some knowledge. So which is it? Ignorance is bliss or knowledge is power? Which would you rather have? Are we doomed to never be happier?

Americans invented the saying "Money can't buy happiness" (and the Beatles knew you Can't Buy Me Love, either). Yet Western culture has consistently chosen money over happiness. And we haven't gotten any happier. Wallis Simpson, the woman who stole Edward VIII's heart and crown, contributed the quintessentially American axiom, "You can't be too rich or too thin." Supermodel Kate Moss chimed in, saying, "Nothing tastes as good as skinny feels." In other words, no matter your personal wealth, there's always the impetus for more—or for your personal weight, there's always the impetus for less. However much you have (or don't), it's just not enough. And therein lies the problem. Because there is no amount of monetary increase or body weight decrease that can activate the serotonin system to provide contentment, especially if you are food restricting. Why? Because money and food trigger our dopamine systems, not serotonin. So we will always want more (or less). Does chasing money bring happiness? And is Kate Moss really happier? Her history of cocaine use belies her $9.2 million salary and mega-stardom.[2] Chasing the wrong dragon. Perhaps that's how she stayed so thin.

We argue that happiness (i.e., contentment) is the goal; everything else, including health and material well-being (money), is the means to get to that goal. So we go to Manolo Blahnik and Tesla and Lululemon and buy the next shiny thing. Even if we don't do yoga, at least we look the part. There's no doubt that material wealth improves subjective assessment of well-being by individuals in the short term. But not in the long term. Is that pleasure or happiness? Is that dopamine or serotonin?

Show Me the Money

Our close personal relationship with money can be summed up like this: "It's not that I want to make more money, I just want to make more money than YOU." We consciously or subconsciously compare ourselves with our peers, keeping up with and wanting to best the Joneses with a pricier house, car, living room furniture, and now drones. The premise is bolstered with a simple experiment. Two decades ago a group of Harvard students were asked a not-so-simple question. Which world would you prefer: a world where you get $50,000 per year while everyone else gets $25,000, or a world where you get $100,000 per year while everyone else gets $250,000? The students overwhelmingly chose the former.[3] They would rather be poorer but better positioned. British economist Sir Richard Layard[4] gleans two findings from this and similar experiments: (1) Your income is judged relative to others. It's not how well you are doing, it's how well you are doing relative to everyone else. But someone is always doing better than you: there are a lot of Warren Buffets out there. *If income is a driver of well-being at all, it is short term, and is not consistent with contentment.* (2) Your current income is judged relative to your previous income. Let's say your salary is X. If you double your salary, 2X is your new income. Next year, even if you make 2X, your well-being next year is the same as you're making X this year. The thrill diminishes. And heaven forbid your 2X salary decreases by Y (where Y is less than X); even though you will make more next year than you did last year, you will be utterly despondent.

This is not how happiness works. Comparing salaries is like taking locker room measurements. Money in the form of income has not translated into individual happiness. We've been programmed to earn more, but then everything in our bracket costs more, and we keep climbing to an unattainable peak. Because income is pleasure, visceral, short-lived, dopamine driven, and subject to all the excesses of tolerance and dependence, and in some cases withdrawal. The cars get bigger, but so does the credit card bill. Americans spend; it's what we do. But it's what you do with the money that determines happiness. Money can often be the means to the end but is rarely the end in itself.

Is *Everybody* Happy?

Does this apply to countries as well? Does material wealth make countries happy? Are countries with higher gross domestic product (GDP) or bigger bank accounts happier than those without? Does GDP translate into happiness? Just as contentment is different from elation, the relation between income and happiness depends on your definition of happiness.

Compared to even fifty years ago, most countries have demonstrated colossal improvements in material conditions, such as clean water, electricity, plumbing, hospitals, and antibiotics to prevent acute infectious diseases. The question is whether these social and medical improvements have improved the happiness of these countries in a significant way. If happiness increases with development, then the enhancement of material well-being should have made human beings and societies happier today than they were at the time of Aristotle. Furthermore, if material wealth were a primary determinant of societal happiness, then those countries with a higher standard of living should manifest more happiness than those with lower standards.

However, the percentage of people who identify themselves as "happy" in terms of per capita income exhibit a tenuous correlation at best, and within the thirty-seven developed countries of the Organization of Economic Co-operation and Development (OECD), not at all. Thus, just like for people, societies are not happier with higher GDP. Known as the

Easterlin paradox,[5] this suggests that, just as people view their income relative to others, countries do the same. Nonetheless, GDP has caught on as a measure of social advancement, to the extent that most countries today are preoccupied with the number, and more than one government has factitiously increased its estimation of GDP to make themselves appear more prosperous.[6]

GDP is defined by the following equation:

GDP = Production + Government + Investment + (Exports − Imports)

A high GDP infers governmental stability, but GDP is subject to manipulation by those same governments. When officials stoke the flames (as in the 2008 economic bailout), GDP can be artificially inflated, but that doesn't mean people are happy. Conversely, if the Fed lowers interest rates, it can spur investment through borrowing, which also artificially raises GDP only by creating a more precarious economic situation. Furthermore, GDP doesn't take into account advances in environmental pollution, or illegal drugs and prostitution, or technology (for good or bad; see Chapter 14). How do you assess GDP in light of the fact that an iPhone is cheaper than its three components (a phone, a camera, and an MP3 player)? It's hard to assess sustainability when car sales are offset by car accidents and car exhaust, yet GDP doesn't account for both sides of the ledger. And it's even harder to assess sustainability and environmental damage when the food industry and the drug industry (which treats the illnesses the food caused) are lumped into one number, thus inflating GDP, while people get sicker and unhappier. It also doesn't take into account unemployment, which is a chief cause of unhappiness. Even economist Simon Kuznets, the originator of the term, in 1929 stated, "The welfare of a nation can scarcely by inferred from a measurement of national income." Indeed, business and government hide behind GDP precisely because it measures productivity exclusive of sustainability. Princeton economist Dirk Philipsen[7] argues that GDP is exactly what is wrong with happiness. Because happiness is long-term, it infers stability and sustainability. GDP is anything but long-term, and it doesn't necessarily improve the happiness of any country's populace.

Recognizing the disconnect between GDP and happiness, and in order to monitor societal advancement or stagnation, social scientists have developed three separate international scales of well-being. The Prosperity Index[8] is a compendium of numerous measures (both in terms of the national economy and personal well-being) that reflect the contentment of populations with their current status. America, for all of its purchasing power, and military and social clout around the world, ranks number eleven out of 142 countries on the Prosperity Index. Given our number one economic standing and quality of life, this is a pretty poor showing. Note that countries with monetary resources don't necessarily score high. Saudi Arabia has oil, Nigeria has diamonds, and they score 45 and 125, respectively. A second scale, the World Happiness Report,[9] takes into account indices that measure the following: real (inflation-corrected) GDP per capita; life expectancy; having someone to count on; perceived freedom to make life choices; freedom from corruption; and individual generosity. Here the U.S. scored number seventeen out of eighty-five countries, and also demonstrated the eleventh-largest drop in the last seven years. A third scale, known as the Happy Planet Index,[10] takes into account only issues of well-being (life satisfaction, longevity, ecological footprint). There, America does even worse, scoring 105 out of 111 countries. So, the data say we're prosperous, but not happy.

The country of Bhutan has embraced the concept that the role of government is to provide a fertile ground for happiness to flourish. They have eschewed GDP as a measure of societal advancement and now utilize the Gross Happiness Index to determine how it is faring as a society.

Bhutan may be a backwater in terms of economic power, but it puts its stock in its people. Perhaps it is because it has a lower standard of living that it is able to focus on the happiness of its citizens. Or perhaps, because Bhutan is a Buddhist nation, it doesn't focus its efforts on the dopamine-driven pathways of its citizens but rather on those that are mediated through serotonin.

Alternatively, you might think that happiness would correlate with economic indicators. You'd be wrong. Take a look at the Prosperity.com website or the U.N. Sustainable Development Solutions Network (SDSN) <worldhappiness.report/>. Right now the five happiest countries are Norway, Switzerland, Iceland, Finland, and Denmark, which was number one. The U.S. was number thirteen. What do they have that we don't? As socialist countries, Norway and Denmark pay very high taxes, and their GDP is one-half of ours. And it's very expensive to live in Oslo and Copenhagen. Food costs in Denmark are almost double that of the U.S. I've been there several times. The restaurants are twice as expensive as those found in America. You go out to the bars on the Strøget, and everyone sits outside and nurses one Carlsberg all evening. Yet, by all these alternative measures, as countries they score much happier. The populace may not have bulging pocketbooks to purchase the extras of life, but they have just what they need to live. And don't really want much more, in part, because each person has similar needs and supports. Also, there is much less economic dichotomy between rich and poor, so internal strife of relative salary disparities is more manageable. Lastly, the costs of basic necessities don't increase faster than their salaries. Norway and Denmark have increased life expectancy, freedom to make life choices, reduced perceptions of corruption, and social support; both countries provide free elementary school, free university, free medical care, and free burial. How can that make them happier? Columbia University economist and head of the SDSN Jeffrey Sachs stated: "There is a very strong message for my country, the United States, which is very rich, has gotten

a lot richer over the last 50 years, but has gotten no happier . . . For a society that just chases money, we are chasing the wrong things. Our social
fabric is deteriorating, social trust is deteriorating, faith in government is
deteriorating."

Alternatively, we might assume that lack of war is a marker of societal
bliss. But, as demonstrated by ancient and recent history, these periods
are few and far between. Perhaps only Switzerland can boast a history
without war. And that's more geographic than historic—because, perched
up in the Alps, they're kind of hard to attack. And there's plenty of
money. Swiss bank accounts are legendary, as they are tax-free, and Switzerland ranks number two on the mean gross income list. And heaven
knows they've got plenty of pleasures to be had, for which they are famous. Switzerland rates close to the top of the Prosperity Index.[11] Considering the U.S. contributes 25 percent of the global GDP, eleventh isn't a
very good showing.

I Can't Get No . . . Satisfaction

These data argue that material wealth is really a reward-driven parame-
ter, not one indicative of level of contentment. Most of the books that have
been written on the relation between money and happiness conflate the
two phenomena of reward and contentment together, into one that is
commonly referred to as "subjective well-being." But is that true? University of Michigan economists Betsey Stevenson and Justin Wolfers have
attempted to discredit the Easterlin paradox on the basis of finding a
logarithmic (i.e., curved) rather than a linear relationship between income and subjective well-being;[12] in other words, each extra dollar is
worth slightly less in happiness than the dollar that came before, but
there is no obvious ceiling. However, I think that these studies suffer

from the same misconception: that, as currently defined, subjective well-being is a meaningful indicator of happiness. How is this question asked of people? There are two ways. The World Values Survey asks, "All things considered, how satisfied are you with your life as a whole these days?" Really? That's supposed to give you an indicator of happiness? The Gallup World Poll asks people to imagine the ladder of life, and which rung they are on. Sounds more like the relative income problem to me.

There is one paper that I think got it right. Instead of lumping, Nobel Prize winners Daniel Kahneman and Angus Deaton examined single individuals rather than in aggregate and, based upon their responses, split happiness into two separate experiences.[13] One phenomenon is equivalent to "life satisfaction," which they describe as "the thoughts that people have about their life when they think about it," such as prosperity and influence (likely mediated through actions that drive dopamine). Using this definition (i.e., at peace with your status in life), one can see a very clear correlation with income, as more money means increased access to services and technology that make life easier (dishwashers, dry cleaners, Amazon Prime). Money can be spent on what matters to the individual, from gadgets to Gucci. However, they also quantified the second phenomenon of contentment, "the frequency and intensity of experiences of joy, stress, sadness, anger, and affection that make one's life pleasant or unpleasant" (equivalent to our definition of *eudemonia*, or our biochemical definition of serotonin effect). Using this definition, contentment demonstrated a logarithmic relationship with income until a maximum of $75,000. After that, the relationship disappeared. Once needs were met, more income did not generate more contentment. It would appear that the acquisition of stuff and property beyond the basics doesn't up your Zen quotient.[14] *The pursuit of property is not the pursuit of happiness.* In fact, it can just leave you wanting more.

Kahneman and Deaton provide us with evidence of how the bio-chemistry plays out in real life: that reward is not contentment, and that increasing reward does not translate into happiness. By pursuing the dollar, Americans have certainly not become content; and by pursuing GDP, America has become derelict in one of its three primary aspirations—*the pursuit of happiness.*

13.

Extreme Makeover–
Washington Edition

Corporations are people. So said the U.S. Supreme Court in the now-infamous decision *Citizens United v. Federal Election Commission* (2010). Well, corporations have a fiduciary responsibility to their stockholders. People have a fiduciary responsibility to themselves and their family. That's not quite the same thing. Also, corporations don't have serotonin, or dopamine, or a Jiminy Cricket. All they have is a balance sheet. This case has been pilloried in the popular press and in public opinion as selling America to the highest bidder. But this case wasn't a fluke. The triumph of corporations over individuals represented by *Citizens United* is the culmination of a forty-five-year-long war fought in the halls of Congress and the aisles of Walmart. Yet, unlike the survivors of a war with weapons, our citizens don't even know that a war was fought, or that it even matters. Corporations now have a legal right to interfere with your pursuit of happiness. The history of the law is just as important as the science in explaining how we got to where we are, and how to move forward—which is why I went to law school.

The Balance of Power

The United States is home to both individual rights and corporate rights. The balance of power between individuals and corporations has always been precarious, one that has exhibited a sinusoidal wave up and down for the first two hundred years of our existence, based on the parity built into the Constitution (thank you, James Madison) and the Bill of Rights (thank you, George Mason!). Since the inception of our nation, corporations have often attempted to tip the scale to usurp control over individual rights, but in each successive era the scale has been rebalanced by the political process. For instance, in the 1870s the robber barons of the steel, oil, and railroad industries and the national banks wielded virtually unlimited power and money. To beat back the threat of monopolization, Congress passed the Sherman Anti-Trust Act in 1890, and in the early 1900s, Teddy Roosevelt was able to effectively constrain the growth in corporations and banks. After the squalor and danger of slaughterhouses and the meat industry were laid bare by Upton Sinclair in *The Jungle* (1906), Congress was pressured to charter the FDA to protect the nation's food supply. After the infamous Triangle Shirtwaist Factory fire of 1911 in lower Manhattan, Congress established the Federal Trade Commission in 1914 to protect consumers and prevent child enslavement. In the 1920s, the next wave of private speculation by companies (such as what happened in the Teapot Dome bribery case) led to "irrational exuberance" and ushered in the Great Depression in 1929. The economic havoc was countered in the 1930s by Franklin Roosevelt's enactment of the New Deal and the Works Progress Administration to get people working again. FDR also established the Federal Deposit Insurance Corporation in 1933 and the Securities and Exchange Commission in 1934 to protect individuals from corporate abuse. World War II in the 1940s and the Korean War in the 1950s, along with Joseph McCarthy's communist witch hunts, rolled back individual rights yet again. Then, in the 1960s, Rachel Carson's *Silent Spring* (1962) exposed corporate contamination of the environment, and Ralph Nader's *Unsafe at Any Speed* (1965) birthed the consumer rights movement. Both books pushed the scale toward the rights of the people, culminating in the establishment of the Environmental Protection Agency and the Occupational Safety and Health Administration in 1970. Has the undulation made you seasick yet?

But the ebb and flow of individual versus corporate power has now ceased. Since 1971 there has been a slow, ever-steady creep of usurpation of power by corporations, with a concomitant loss of power from individuals. Look, I am not espousing socialism or communism or any other "-ism." But the balance of power has so shifted in favor of the rights of corporations that individuals are losing, and in ways that are opaque to them. They have more consumer choice, so they think they have more rights, but in fact they have far fewer. This shift in the balance of power can in part be traced, as documented by City College sociologist Nicholas Freudenberg,[1] to the tenure of one particular American—U.S. Supreme Court associate justice Lewis F. Powell Jr.

Corporate "Justice"

Why would a Supreme Court justice be the thumb on the scale that tipped America to favor corporate over individual interests, on which so much of our current unhappiness is grounded? (See Chapter 14.) How does one man out of nine wield that much power, and in the judicial branch, no less? Does the Bill of Rights count for nothing? Powell's biography gives us several clues as to his philosophy and his methods. In

World War II, Powell worked in U.S. counterintelligence, where he learned the need for absolute secrecy. Following the war, for twenty-five years he was a partner at Hunton, Williams, Gay, Powell and Gibson, a Virginia firm specializing in litigation and business law. During this time, he also served as chairman of the Richmond School Board and advocated "constant surveillance" of school textbook and television content. He was at the helm of that school board when the Commonwealth of Virginia defied the Supreme Court's mandate for integration in *Brown v. Board of Education* (1954). He also served as the liaison between Virginia Commonwealth University and the tobacco industry. The first salvo sounded in 1950 with the now infamous Doll paper[2] equating smoking with lung cancer. At that point Lewis Powell's firm began to provide legal defense for Big Tobacco. From 1964 through 1971, he served on the board of directors of tobacco giant Philip Morris. In fact, he represented the Tobacco Institute and various tobacco companies in numerous law cases, defending them against individual and class action lawsuits for lung cancer. In 1971, because of his corporate ties and pro-business leanings, he was appointed to the Supreme Court by President Richard Nixon. Note that Justice Powell was never a state or federal judge nor was he a constitutional lawyer.

In 1971, prior to his nomination for the high court, Powell wrote a secret memorandum to his friend Eugene Sydnor, who was the chair of the U.S. Chamber of Commerce. In this now-notorious document, entitled "Attack on the American Free Enterprise System,"[3] he bemoaned the loss of American standing in the world in the 1960s due to the civil rights 'and counterculture revolutions, as well as decrying the loss of corporate power due to the ongoing assault of both "extremists of the left" and "perfectly respectable elements of society." In the secret Powell Memorandum, he admonished Sydnor to help take back America from the hands of the mob. Corporate America, he stated, must become more aggressive in influencing and molding politics and law. He argued that the U.S.

Chamber of Commerce, at that time a passive pro-business lobby group, must itself become politically active. He stated that business must not waver and take on prominent bastions of public opinion and political power—the universities and the courts.

Not only did this resonate with Sydnor, but this "confidential" Powell Memorandum quickly spread throughout the business community—not quite viral, but more parasitic. This missive can be traced to the origins of several right-wing think tanks and lobbying organizations, such as the Heritage Foundation (the origin of Reagan's *Star Wars* missile defense system) and the American Legislative Exchange Council (a front group for the food and pharma industries). By the time this document revealing Powell's biases and motives was leaked to muckraking *Washington Post* reporter Jack Anderson, who reported it in 1974, Powell was firmly ensconced on the Supreme Court.

Like a worm that slowly eats away one apple at a time in the barrel until the entire barrel is contaminated, Powell's action on the Supreme Court ate away at individual rights, with the result that we now have diminished individual contentment and increased societal anxiety. Let's look at some of the seminal cases where Powell voted with the majority.

- 1976: *Va. State Pharmacy Board v. Va. Citizens Consumer Council.* This case allowed unfettered corporate advertising to unsuspecting patients, paving the way for all the drug ads you now see on TV. This case has helped the pharmaceutical industry harness fear to create demand for their products (see Chapter 14).
- 1976: *Buckley v. Valeo.* This case was the precursor to *Citizens United.* The decision did away with limitations on campaign spending and individual donations in elections (thirty-four years later, *Citizens United* allowed corporations to turn their treasuries into campaign war chests to bankroll any candidate to do their bidding). Now public elections are a free-for-all, where politicians,

high-rolling donors, and corporations can buy elections, and the most money wins. Higher betting means higher pay-outs. And, to quote Will Rogers, "A fool and his money are soon elected."

- 1978: *First National Bank of Boston v. Bellotti.* This was a 5–4 decision, and also set the groundwork that paved the way to *Citizens United.* Powell was the deciding vote and wrote the majority opinion. This case basically said that corporations could say whatever they wanted and vote with their dollars. Of course they do say whatever they want, and they vote for themselves in the form of campaign contributions. When there are more dollars than votes, it becomes a societal problem.

- 1980: *Central Hudson Gas & Electric Corp. v. Public Service Commission.* This was another 5–4 decision, with the majority opinion authored by Powell. Up until this case, public utilities were just that: public. They could not have an opinion and they could not advertise, because they were a public trust. Powell saw it differently. By the time Powell got through, he had protected "commercial speech from unwanted governmental regulation." He instituted a four-part test for government to regulate commercial speech: (1) Does it violate the First Amendment? Well, only speech that breaks the law (incitement to riot, yelling "Fire!" in a crowded building) isn't covered. Otherwise it's fair game, even if it is false. (2) Is the government interest substantial? That means that short of trying to overthrow the government, anything goes. (3) Does the regulation advance the government interest? Corporations could sell poison to the public, but as long as people aren't dying in the streets (i.e., obvious deception), government stays quiet. Tobacco was a good example. And finally: (4) Is the regulation as narrowly tailored as it can be? In other words, there must be a "reasonable fit" between the government's ends and the means for achieving those ends. Oh, by the way, it's not *any* of the four; it's got to be *all four* that are violated at once. So virtually any speech can be corporate speech.

In the span of four short years, 1976 to 1980, Lewis Powell helped to lay waste to any hope of keeping corporate power and influence in check. If you want an analogy of the degree to which one man can change the landscape of America, think back to Frank Capra's timeless Christmas classic *It's a Wonderful Life* (1946). Jimmy Stewart plays George Bailey, the forlorn do-good head of the Bedford Falls Savings and Loan, who holds the mortgages of virtually every townsperson and who, on the brink of committing suicide, is visited by a Christmas angel who shows him what his hometown would look like if he'd never been born. A very different place where prostitution, alcohol, and guns are the fabric of society. Pleasure and reward is primary, to the delight of the bankers and to the detriment of the citizens. Indeed, we're all now living in Potterville . . . I mean, *Powellville*. All dopamine all the time, with a soupçon of cortisol thrown in to stir the pot.

The Rights of the "People"

This is how the U.S. Supreme Court has paved the way for the relentless marketing of products, irrespective of their utility or cost, either personal or societal (e.g., Big Pharma, Big Food). And there's virtually nothing left in the public legal arsenal to curb their influence. On top of that, we have now elected the first "populist" president, who is on the side of corporations versus the government instead of on the side of the people. The marketing salvos come swift and hard, they're aimed at our nucleus accumbens (food), our amygdala (drugs), and our prefrontal cortex (our Jiminy Cricket), and we can't possibly defend against all of them.

Corporations are now blessed with the ability to produce and subject us to virtually any unfettered form of corporate advertising, campaign spending, and commercial speech. Not only that, but because of Justice Powell's legacy, corporations have even more rights than people do. They have both the rights of corporations *and* the rights of individuals. Nobel Laureate Joseph Stiglitz chronicled the fact that corporations can now sue the federal government for restitution of profit lost from the enactment of any government policy.[4] *You* can't sue the government, but *they* can? Are you kidding me? And there are no prosecutions of individual lawbreakers, as they derive protection from civil actions by their position as an instrument of the corporation, and you can't lock up a corporation. Witness the Economic Recession of 2008, perpetrated by Big Banking against the public using the vehicle of subprime mortgages. Lehman Brothers, Washington Mutual, Countrywide—all gone, and their investors' money with it. Merrill Lynch was forced to merge with Bank of America, and Bear Stearns with JPMorgan Chase. AIG and Citigroup survived only because of the government bailout. Lots of stress and cortisol and very little serotonin. But who was prosecuted? A total of one person—a Credit Suisse banker, Kareem Serageldin, born in Egypt. The judge who sentenced him called him "a small piece of an overall evil climate within the bank and with many other banks."

Why didn't they prosecute the rest of these corporate hoodlums? Simple. They work for corporations. If corporations are "people," then do you employ the rules regarding corporations, or do you employ the rules regarding people? Do you call it a lack of caution, in which case it is a corporate civil action with no jail time? Or do you call it personal "criminal fraud," a felony punishable by incarceration? The Department of Justice has chosen the former, in part because the Court's current leanings would undo any such criminal case.

Given their pro-business behavior, you could argue that the success of the Corporate Consumption Complex versus the American public lies squarely at the doorstep of the Supreme Court. But that would only be half the story. Congress has also played a significant supporting role in creating our current reward-driven culture, which allows for marketing of virtually anything and everything (see Chapter 14). Nowhere is this better exemplified than in the story of how Congress changed its tune in the span of just one decade around the promotion and marketing of junk food (see *Fat Chance*), aimed at one target: dopamine.

When Legislature Is Corrupted, the People Are Undone

The junk food industry started to take off in the 1970s in part due to (1) improved harvesting and food processing technologies by companies like Archer Daniels Midland and Cargill, and (2) the enormous federal subsidies for corn, wheat, soy, and sugar supplied to processed food companies as part of the Farm Bill, which made them cheap. The original Farm Bill, which dates to 1933, had as part of its legislation the Agricultural Adjustment Act of 1933 (revised in 1938), which paid farmers not to grow certain crops and to kill off certain livestock—this effectively reduced supply and artificially increased prices—and farmers got to double-dip. This policy diametrically changed course in 1971, when President Richard Nixon decided that fluctuating food prices caused political unrest (and he had a lot of unrest to deal with). He remanded his agriculture secretary, Earl Butz, to make food cheap. Butz immediately canceled the government handout, and said three things to the American farmer: "Row to row," "Furrow to furrow," and "Get big or get out." Government now would reimburse on quantity, not quality, and the farmer would make it up in volume. Let's use corn as an example. If you make corn more plentiful, supply and demand says you sell it for less. But if some of that corn is turned into high-fructose corn syrup, and that's put in every processed food, then you can sell it for more. And if you turn some of that corn into ethanol, you can sell it for even more.

Also in the late 1960s and early 1970s, just as this massive agricultural overhaul was beginning, the sugar industry came under intense scrutiny due to nutritional research performed by John Yudkin in the UK and Sheldon Reiser in the U.S., which correlated sugar consumption with heart disease. In response, the Sugar Association established a public relations arm called Sugar Information, Inc., and hired a PR firm to blanket the country with pro-sugar propaganda. Saying that sugar was the "quick energy" to provide "the willpower you need to undereat" and "to curb your appetite"—in other words, deceptive advertising designed to drive your dopamine upward. In 1972 the Federal Trade Commission took Sugar Information to court[5] and shut it down. This was heralded as a major win for consumer watchdog groups and demonstrated the power of government to protect the public.

Unfortunately, this episode was also the immediate impetus for the chartering of the American Legislative Exchange Council (ALEC) in 1973. ALEC's website states that "Americans deserve an efficient, effective, and accountable government that puts the people in control" (www .alec.org). But it doesn't define which people. What ALEC does is draft "model legislation" that favors its funders, and then feeds these model bills to state or federal legislators to get them passed. A new way to perform congressional lobbying. A corporate bill mill and front group. When you hear a politician railing against corporate special interests, they're talking about ALEC. And who are the funders? Well, there are a lot of them, but the food and pharma industries were founding members, and agriculture, pharma, and guns are numbers one, three, and four on their bill-writing agenda. ALEC doesn't just write the bills, they provide campaign contributions to the legislators who introduce them into law, which currently totals 338 out of 535 congresspeople.[6]

Special interests grew exponentially in the 1970s, especially after the

Supreme Court verdict in *Buckley v. Valeo.* In 1970 there were 175 lobby-ists in Washington, D.C.; by 1980 there were 2,500. In 1970 there were 300 established political action committees; by 1980 there were 1,200.

This leaves us with a conundrum: Who's in charge? We pride our-selves on being a democracy. We choose who to vote for, and what we want to spend our hard-earned money on. Or do we? When corporations can act with impunity, unbridled, without oversight, and both dictate and cater to our deepest wants and desires, where is the accountability? This is the model theorized by Princeton philosopher Sheldon Wolin, who warned us of the evolution of a political hybrid in which corporate America and Washington were one and the same.[7] Political scientist Martin Gilens goes even further[8] by showing that economic elites and organized groups representing business interests have substantial inde-pendent impacts on U.S. government policy. For instance, if a congres-sional bill is supported by 20 percent of the rich, it passes 20 percent of the time; whereas if a similar bill is supported by 80 percent of the rich, it passes half the time. As long as America views the concept of happiness as consistent with both consumption and GDP, it will not be able to break free of corporations' stranglehold on our brains. And they know what they're doing, because they've done the research and know what works. Read on, and you will too.

14.

Are You "Lovin' It"?
Or "Liking It"?

Although corporations are now people, they are not faceless behemoth monsters. They are comprised of fine and upstanding individuals, each with principles, morals, and high aspirations. But they work for a corporation, whose only job, in the words of Goldman Sachs CEO Lloyd Blankfein, is to make money. They are very good at what they do. In the process they have honed the ability to exploit human emotion. It is called *marketing*. By conflating the notions of pleasure and happiness, they know how to get a rise out of your dopamine and cortisol.

Many Happy Returns

My personal favorite is Coca-Cola's "Open Happiness," first unveiled in January 2009 and still going strong. This is the longest-running media campaign for the 131-year-old company, which has arguably mastered the messaging of its own concept of happiness to the masses. So if this is the longest campaign, you know it must be working, both within the U.S. and worldwide. In 2015, Coca-Cola aired a commercial in Mexico featuring a group of white hipsters who drove into a Mexican village bearing housing supplies and coolers of Coke. The villagers smiled with what appeared to be genuine happiness at the entrance of these well-meaning outsiders and their ability to construct a Christmas tree made out of Coca-Cola bottle caps. The commercial was eventually pulled but was exemplary of capitalizing on America being the ideal and Coke being the way to get there.

Let's see, high-fructose corn syrup (or sucrose in Mexican Coke)—that increases dopamine. Caffeine—that increases dopamine. Water, phosphoric acid, carbonation, salt, caramel color . . . I'm not seeing it. And the metabolic syndrome that results from continued excessive sugar consumption—that decreases serotonin. A bottle of Coca-Cola can certainly provide a little reward, but it is a far cry from contentment. Coca-Cola isn't the only culprit in marketing dopamine and disease to the masses through sugar.

How about "I'm Lovin' It" from McDonald's? This was the company's tagline from 2003 until 2015. They veered off course in 2012 by tipping their hand with something a little too close to home: "Crafted for Your Craving"—in essence admitting to consumers that they have addictive intentions in mind. But that campaign didn't last long. McDonald's sales are currently flagging in part due to international consternation over the overwhelming links between processed food, obesity, and diabetes. In 2015 they fired their CEO Don Thompson due to flagging sales. Do you think it's called a "Happy Meal" by accident? That they pitched their product to children with plastic trinkets and smiling characters accompanying the joy of sugar and fat and salt? That dopamine/reward surge starts early, and they're branding kids for life. Every day, McDonald's serves up 3.2 million Happy Meals, grossing more that $10 million per day, and kids who order them now account for 14.6 percent of customers. McDonald's is pleasure (although that is highly debatable). They've been called out on their role in the obesity epidemic, and they are now attempting an about-face, including options of apple slices (don't forget the caramel dipping sauce—scrapped in 2011 because of the obesity epidemic). But most kids still prefer the fries. And chocolate milk has just as much sugar as the soda. If you're taking the kids to McDonald's for a Happy Meal, I doubt you're ordering the salad instead of the Big Mac.

Perhaps the most egregious product placement to confuse the public is the one-hour viral YouTube video (now at 56 million views) *Happy (12AM)*[1] posted by Pharrell Williams's foundation using his song as the soundtrack. Within the first two minutes of the video, Pharrell enters a gas station convenience store at midnight and purchases two candy bars, a bag of potato chips, and a Red Bull, then offers them up to the audience, singing, "Clap along if you feel that happiness is the truth." First of all, while pleasure can be found in a candy bar (double your pleasure with two?), happiness is nowhere in sight. And second, if you're drinking Red Bull at midnight, your sleep deprivation with its concomitant cortisol rise will put you at risk for some big-time unhappiness down the road. Clearly, Pharrell's view of happiness is colored by corporate sponsorship of his non-profit.

What about the iconic happy hour? Is this pleasure or happiness? Which brings the customers into the restaurant—the $5 pupus, or the $5 drinks? Across the board, alcohol has grown as a percentage of sales from 9 to 15 percent, and in sports bars alcohol makes up 26 percent of sales. Alcohol has always been viewed as a sales lubricant for restaurants, even during a recession, because it's the pleasure the public can still afford. But what about when that happy hour is at 10:00 a.m., and televised? NBC has broken all the rules about drinking on TV, and the fourth hour of the daily talk show is now dubbed the "TODAY Happy Hour." Two

middle-aged women drinking from large margarita glasses in the morning. This hour even won a Daytime Emmy Award. This would be funny
except that women are 40 percent more likely to develop anxiety than
men are and often turn to alcohol to cope. Hoda Kotb, the show's cohost,
says, "It's a joke. The producers just keep coming up with ways of getting
us to drink, it's become our thing. It's like a nonstop party from ten to
eleven."[2] While this may seem lighthearted, the network is normalizing a
behavior that could have serious health implications (e.g., breast cancer)[3]
for women, who are the target audience.

Existential T(h)reat

And now for something completely different. Imagine sitting naked in a
bathtub on a cliff overlooking the ocean at sunset, with your partner in
an adjacent bathtub. The dopamine surges in motivation for anticipated
pleasure. Then, overwhelming fear and angst with the possibility of erec-
tile dysfunction (ED). By introducing fear as part of the marketing campaign, Viagra, Cialis, and Levitra have all kicked it up a notch; ED drugs
are expected to account for $3.2 billion in global sales by 2022, with more
than half of those sales in the United States. There are certainly men who
suffer from ED, and there are countless more who buy the product based
on fear.

Fear has been a primary driver of consumption since the inception of
marketing. It started with car dealers and high-pressure sales tactics.
Since then, virtually everything from mouthwashes to dishwashers to
Hummers to Smith & Wessons are sold out of fear—either "fear of failure" or "fear of the unknown." Overstock.com and Groupon are examples of driving sales through the fear that you might miss out. By combining
a short-duration offering with the threat of scarcity and the "keeping up
with the Joneses" factor, marketing makes sure that fear remains front

and center. Because fear means stress, and stress means cortisol, and prefrontal cortex be damned, it's time for the chocolate cake.

Strategies That Sell

Is this marketing or propaganda? Definition of marketing: the action or business of promoting and selling products or services, including market research and advertising. Definition of propaganda: information, especially of a biased or misleading nature, used to promote or publicize a particular political cause or point of view. Since pleasure and happiness are clearly not the same thing, the conflation of the two is inherently biased and misleading. Therefore, advertising that implies that the selling of reward as contentment is by its very nature propaganda. Pleasure from hedonic substances and pharmaceuticals (masquerading as happiness) can be easily purchased. If you don't know the difference between the two, it stands to reason you will lay your money down, and then they've got you. Like the pusher on the playground who gives you your first free hit, they'll turn you into a customer for life.

Old-style marketing (direct and telemarketing) was across-the-board and hit-and-miss, based on demographics, unsolicited contact, and the ability to generate fear in the consumer. With the advent of the internet, marketers honed their messaging to specific groups based on their previous "likes" and searches. And now the new discipline of *neuromarketing* is taking the guesswork out of the equation and increasing efficacy of sales. In neuromarketing, the brain responses of subjects to industry messaging are analyzed. This allows those companies to hone their messages to specific subgroups within the larger masses and generate even greater profits. It's now public knowledge that Coca-Cola will use neuromarketing in all quantitative ad performance projects in the coming year to "spread happiness." According to branding agency Kantar Millward

Brown,[4] facial coding will be the primary technique used to gauge consumer emotions. The technology is seamlessly integrated: they record the subjects' faces while they watch ads within a normal survey environment, automatically interpreting their emotional and cognitive states moment by moment. Facial coding was originally the province of experts, who viewed slow-motion video of subjects to record fleeting "true" emotions that register briefly in facial expressions. Kantar Millward Brown's system uses eye tracking and other phenomena to measure engagement, brand association, and motivation, among other metrics. And they use these data to target . . . you. Unilever (the conglomerate that owns Dove soap, Lipton tea, and Ben & Jerry's) is also pursuing a similar 100 percent testing approach. If this sounds Orwellian, it is. And it's here. And it works to drive dopamine and cortisol, in a pitch to get you to buy more. The problem is the more you buy, the unhappier you get.

The purveyors of hedonic behaviors, devices, and consumables are all looking for that winning formula to provide the public with some form of product (requiring continued purchase), along with an inherent hook that will maintain or even increase consumption and in which the market never reaches saturation to allow for continued expansion. Marketing genius Nir Eyal provides a window into the hedonic platform used by companies to hook us and keep us coming back for more.[5] According to Eyal, every successful product consists of four intertwined concepts that drive an unending vicious cycle. (1) The *trigger*; that is, something that commands your attention even when you don't want it to, like an itch. (2) The *action*; that is, a stereotyped behavior that somehow soothes the trigger, which is easy to perform, does not require thinking, and can be accomplished in mixed company. In other words, a scratch. For instance, clicking on your e-mail or Facebook account is easy to do, does not require thinking, and is currently socially acceptable. Conversely, soothing an incipient sexual urge may depend on the venue. (3) *Variable reward*, the most important part of the cycle. These can be social validation rewards like Facebook or Instagram; intrinsic motivation rewards such as points scored on video games; or sustenance motivation rewards (e.g., money or calories burned) in video poker or MyFitnessPal. If the variable reward is the result of a behavior, then it is the *inconsistency* of the reward that ultimately drives that behavior to become a habit. (4) The culmination is *investment*, which is really the only thing that drives company sales. We internally rationalize why we needed this reward in the first place (even though the reward was variable, and even though we had previously lived just fine without it), and that the cost of the product becomes well worth the new habit because we can now soothe the market-generated itch in a culturally acceptable manner.

A "Slot Machine" in Your Pocket

Neuromarketing is just one of many new technologies designed to amp up the dopamine to increase sales, but with the unintended consequence of making us miserable. Drunk driving, though still a common practice, is taboo. People utilize designated drivers or Uber. Unfortunately, society hasn't bought into the concept of using a designated driver if you're a compulsive texter. In 2006, a nineteen-year-old student engaged in texting while driving on an otherwise deserted road in Utah killed two astrophysicists in a head-on collision. Since then, fourteen states have banned the practice, yet it goes on unabated. MADD (Mothers Against Drunk Driving) has yet to form a MACT (Mothers Against Compulsive Texting). But the death and accident rates are becoming increasingly similar.

Apparently, the draw of the screen is just too much for most people; the cell phone is like a slot machine. With every ding, a variable reward, either good or bad, is in store for the user—the ultimate dopamine rush. As Robert Kolker wrote in the *New York Times Magazine*, "Distraction is the devil in your ear—not always the result of an attention deficit, but borne of our own desires." We are distracted because we want to be. Because it's fun and obfuscates real life. Why else would they sell so many smartphones? My wife says that I'm addicted to my e-mail, and I know looking at it doesn't improve my mood. A good gadget is essentially a wondrous object, commandeering our focus with delight and surprise (Steve Jobs used the word "magical" about the iPhone when it debuted). The smartphone brilliantly exploits two types of attention: "top down" (what we want to focus on) and "bottom up" (what takes us by surprise).

That need for surprise is what it's all about. Surprise is visceral and immediate, and stokes our dopamine and our nucleus accumbens. But it's fleeting, and rarely does any happiness come out of it. In fact, the frequent checking of cell phones, waiting for something to change, is linked to anxiety and depression.[6] Of course, again, correlation is not causation. Do cell phones cause depression? Or are depressed people trying to eke out a little dopamine rush? Or both? I'll tell you one thing: cell phones certainly don't bring serenity.

Mobile Madness

Does cell phone use drive cortisol, the other bad boy in this paradigm? Cell phone use is linked with stress, sleep loss, and depression in young adults (although of course causation cannot be proven). A recent study in young adults showed that cell phone use was negatively linked to grade point average—the higher the cell phone use, the poorer the grades.[7] They also found, perhaps unsurprisingly, that higher GPAs tended to correlate with more happiness, while more anxiety was linked to less happiness. Anxiety and happiness were assessed with two well-known questionnaires for assessing mental health: the Beck Anxiety Inventory and the Satisfaction with Life (SWL) index. Statistical analysis on these associations encouraged the researchers to suggest cell phone use is linked—via GPA and anxiety—to loss of happiness. Another study demonstrated that fourth and seventh graders who sleep with cell phones in their room get less sleep than those who don't,[8] although we can't say whether they're playing games on them or if the problem is just the glow of the screen. We do know that sleep deprivation increases food intake and risk of weight gain (see Chapter 9), driving further unhappiness. In a tragic example of distraction by technology, a South Korean couple obsessed with raising their two "virtual children" online let their actual three-month old daughter starve to death.[9] It's not only affecting teenagers. Rehabs are popping up treating "device addiction." There have been reported cases of withdrawal. While opioids get the most press, internet and gaming addiction is leading to social devolution in large numbers.

From World of Warcraft to Call of Duty to Pokémon Go, video games have been linked to bingeing, and even a few cases of excessive sleep deprivation resulting in death. The Chinese have noted white matter changes in teens and young adults who binge on the internet, and have labeled this phenomenon Internet Addiction Disorder.[10] But is this really addiction (see Chapter 5)? Internet and gaming disorders are not yet sanctioned as valid psychiatric diagnoses but are now being considered. In these behavioral addictions, both the nucleus accumbens (NA) and the prefrontal cortex (PFC) are severely dysfunctional.[11] Does this lead to depression? One study tracked teens in alternative high schools for one year post-graduation. Those who exhibited anhedonia (difficulty experiencing pleasure) at baseline were more likely to indulge in internet game bingeing and to manifest signs of depression one year later.[12] So which came first, the video gaming or the anhedonia? Were these the same kids who'd be listening to Depeche Mode and wearing Goth clothes thirty years ago? Are students who graduate from alternative schools already self-selected for behavior problems?

The Bully Pulpit

Smartphones have ushered in yet another method of misery for adolescents. School bullying dates back to the advent of organized school in the eighteenth century. Bullies always wielded an advantage over the bullied, which imbued them with a sense of perceived authority. These advantages could be physical, such as size, gender, age, or weight; or social, such as clothing, cliques, or academic status. Nearly 160,000 kids stay home from school every day because of fear of bullying. Bullying has always been an issue, but more and more schools have adopted no-bullying policies. Thus, the bullies have gone underground. Cyber-bullying, the newest hip way to express rage, has become all the rage. Nowadays, you rarely hear about the bloody nose—rather, you hear about the suicide. More than one in three young people have experienced cyber-threats online, one in four have been bullied repeatedly online, and over half of adolescents have participated in some fashion. Yet most young people do not tell their parents when cyber-bullying occurs. While teenagers have always been spiteful, the ability to hurt people online while maintaining a geographic distance has made cyber-bullying rampant. One Florida twelve-year-old girl was terrorized by as many as fifteen girls who picked on her for months through online message boards and texts. One message said she should "drink bleach and die." Instead, she jumped off a cement factory tower.

"Like" Sitting Ducks . . .

In particular, digital technology has created a relatively new and ubiquitous form of psychological stress. Do you spend time on Facebook or any type of social media? It's become the social norm to act/comment without thinking, posting the newest inflammatory memes or tweets, without considering context. We're becoming more of an immediate-gratification (dopamine-driven), knee-jerk reaction (PFC inhibition) society—making our lives about the number of likes we can receive.

This is especially true in adolescents and young adults. At the extreme, teenage digital media abusers exhibit the "descent into Hades" (Chapter 5); interestingly, for boys it's video game addiction, while girls instead appear to suffer from social media addiction.[13] While not all scientists agree on the criteria for these disorders, and whether these qualify as their own disease processes,[14] there is more than enough data to demonstrate a correlation between internet use and depression. For adolescents who perceived that they had few friends, internet use for communication (e.g., texting) provided "some" form of communication; while for those with no friends, internet use for non-communication purposes (e.g., "surfing") predicted more depression and more social anxiety over time.[15] Maybe because they spent all their time comparing themselves to an elusive ideal, or staring at the photos of their peers going to parties they weren't invited to?

The "Like" button made Facebook the most accessed website on the entire internet. That "Like" button wields more power than virtually any fist, but new data suggests it damages both the Liker and the Likee. Girls in particular post selfies, waiting in anticipation for the Likes to roll in. When they don't, there's an obvious problem in social standing. Although Facebook isn't as hip now as it was a few years ago among adolescents, the newer sites, like Snapchat and Instagram, basically do the same thing. One recent study demonstrated association between the use of Facebook and the development of depression, but only in those teen girls who used Facebook as a surveillance tool to compare themselves to others, which, realistically, is most of the adolescent population.[16] (More on Facebook and social media in Chapter 16.) So is this cause or effect? If you're an insecure teen already predisposed to depression, you might be prowling the internet to see what and who other kids are Liking. Your cortisol is already doing a number on your PFC, your serotonin receptors are already diminished—but the "Like" button takes teenage angst and misery to new heights (or depths). Adolescence is a painful time. Cell phones and the internet may encourage networking and creativity, but it comes with a cost.

In a seven-year follow-up study of over four thousand teenagers, total media use correlated with the prevalence of eventual depression, especially in boys.[17] More media use meant greater risk for depression. One major question that these data continue to pose is the issue of cause or effect; in other words, does internet exposure lead to depression, or do kids with risk for depression rely on the internet as an outlet for their anxiety and for self-expression? It's a pretty difficult issue to prove, and we don't know for sure, especially with how quickly technology changes. But in human studies, the nucleus accumbens and the PFC show characteristic changes with excessive internet use[18] (see Chapter 14), suggesting that the enhanced motivational value and uncontrolled behavior is being driven by structural changes in the brain, which are themselves driven by internet use. AND THAT IS WHY WE MUST EXPRESS OURSELVES IN ALL CAPS WHEN WE WANT TO BE HEARD!!!!!!!!

"Like"-wise, the Liker can get into trouble. A recent study created a fictitious social media group, in which all the participants were actually the research subjects, and the scientists were in charge of what each subject viewed on Facebook.[19] Each subject underwent MRI scanning while being shown a fake item for a thumbs-up or thumbs-down. The nucleus accumbens (NA) lit up only when the subjects Liked something that they thought that others Liked as well—in other words, they exhibited the herd mentality. If they Liked something that wasn't popular, no dopamine rush was noted. This could easily be a trap for both participants on the end of a social network. Maybe we should call it antisocial media.

And Facebook, Snapchat, etc., are for-profit entities. They are there to make money through ad sales. They need to be relevant to do so.

Irrational Exuberance

If you think your smartphone provides a pretty good rush, try the floor of the New York Stock Exchange. John Coates is a Wall Street trader turned University of Cambridge neuroscientist who studies traders on the floor of the stock exchange to determine when, how, and why they engage in risk-taking behavior.[20] He chronicles both the testosterone and the dopamine surges of the traders in the midst of a bull market versus the cortisol rises during the bare-knuckled fall to the bottom of a bear market. The result of this double hit is an overworked and overtired reward system, unable to muster up a taste of victory. Perhaps then we should not be too surprised to hear that David Nutt, the UK drug czar, opined in 2013 that many of the traders who precipitated the Great Recession were so morally bankrupt, so depleted of dopamine and their receptors, that they had to resort to cocaine to feel *anything*.[21]

No doubt, conflating happiness with the maximization of pleasure and reduction of pain (see Chapter 1) has influenced the inherent structure and function of our financial markets. In part due to the Great Depression, the Keynesian economic school of thought ruled the markets from 1936 until 1970. Keynesian economics states that the private sector makes the decisions, but always under the watchful eye of government, which alters policy as necessary (i.e., establishes regulations). Such an oversight posture invariably limits growth and therefore the production of money. Rather, Milton Friedman[22] and his University of Chicago economic colleagues conflated money with happiness, and his legacy is that more happiness, i.e., more money, is always better than less. After all, consumers want more *bang* for less *buck*, and more *buck* for less *bang*.

Indeed, the genius of the Chicago school was to apply this psychology known as "price theory" to all walks of life—that the only method of rational behavior was that which created the most happiness . . . er, money.

Enter Lewis Powell. By 1980 the sands had shifted to favor corporations over people. The Chicago school began to dominate in 1980 under Reagan. Banks started borrowing at low rates to buy other companies to liquidate them, known as risk arbitrage. The Chicago school achieved its sentinel victory with the repeal of the Glass-Steagall Act in 1999, which completely deregulated the banks and markets—and we all know what came next.

What the Market Will Bear

Yet for all their unpredictability and volatility, markets still do work, and we usually let them. Except when it comes to addictive substances. Witness the phenomenon of price elasticity for foods.[23] This is an index, applied to products, that indicates how badly the consumer wants the product using the metric of how much they would be willing to pay, and it is driven by dopamine and its receptor.[24] The price elasticity index is measured as: if the price were to increase 1 percent, how much continued sales would remain? A low index means that people stop buying the product, and the product is "price elastic." A high index means that people will continue to buy the product even though the price has increased, thus the product is "price inelastic." The most price elastic food item is eggs, at 0.32. This means if the price of eggs goes up 1 percent, consumption goes down 0.68 percent. Eggs are the highest-quality protein there is. Eggs have all the nutrients you need. They are literally the world's most perfect food. And people won't buy them if the price increases. Why? Because there's nothing in an egg that has hedonic properties. Tryptophan (the precursor of serotonin) sure, but can it drive dopamine? Conversely, the most price inelastic consumable is fast food, at 0.81. This means if the price of fast food goes up 1 percent, consumption only goes down 0.19 percent. And the second most? Soft drinks, at 0.79. These two food items exert the most hedonic effects (due to sugar and caffeine) and happen to be the ones that people will consume no matter what. And of course they are the most addictive. So how can society turn an addicted, depressed, drug-addled, corpulent, and metabolically ill populace around?

Each of these hedonic stimuli (substances or behaviors) generates money for its purveyors or they wouldn't be purveying it. But what does it do for the individual and for society? In 2007 the U.S. gaming industry generated $92 billion of which about 15 percent was profit. Although 86 percent of Americans gamble once a year, only 16 percent are frequent gamblers, scratching the lottery ticket at least once a week. An estimated 36 percent of the revenue comes from "problem" (i.e., addicted) gamblers, which is 1.1 percent of the population (2.2 million people). So when you do the math, the addicted gamblers lost $33 billion compared with a $14 billion profit to the gaming industry. More money lost out of the system than gained. But no one cares about this, because it's "pay to play"; if you lose, it's your fault. And 2.2 million is not that many people.

Data on the alcohol industry is harder to come by, but one report says that alcohol revenues in 2013 were $308 billion. Twenty billion dollars was profit due to the high taxes on alcohol. But only 61 percent of the population imbibe alcohol, and 20 percent of America are problem drinkers (9.6 million) who spend $3,200 per person annually. So $30 billion of the revenue is spent by alcoholics. And the rehab industry can gross $50,000 per month per suffering addict. Again, more money lost out of the system than gained. Similarly, although 10 million people are alcoholics, most people don't think this is a big problem, because (1) unless you have an alcoholic in the family, it's not your problem; (2) if you drink, it's your fault; and (3) alcohol is already regulated.

Now let's look at the food industry. The food industry grosses $1.46 trillion a year, of which 45 percent is gross profit (which makes gambling and tobacco and alcohol look like chump change). Yet the U.S. health care system in 2015 spent $3.2 trillion per year,[25] of which 75 percent are diet-related chronic metabolic diseases, and of those, 75 percent could be prevented. That means $1.8 trillion a year being wasted on preventable health care—triple the money than the industry profits. And now, since we're talking about more than 50 percent of the U.S. population afflicted with some form of chronic metabolic disease (driven by the sugar in processed food), this is an enormous problem. And it's not their fault, because they didn't pay to play; the sugar was put there by the industry for its own purposes. And since the amount of money per family wasted on health care (18 percent of GDP) is way above their food budget (7 percent of GDP), this constitutes a policy crisis as well.

This is tantamount to what has happened to the health insurance industry. For decades, health insurance followed the casino model: (1) pay to play, and (2) set your own rates. In this model, the insurance industry wanted people to be sick, and they couldn't lose. And as long as there was enough money in the till, they pulled down big profits. Obamacare put 32 million sick people onto the rolls, and they were capped at 15 percent profit. We'll see what Trumpcare looks like, but the proposed rollback of regulations is unlikely to improve health or health care or insurers. The one thing we know for sure is that insurance companies now want their subscribers to be healthy, because they can't make as much money off sick people. But the populace is sicker than ever, in part due to chronic metabolic disease and in part due to mental health conditions, neither of which will improve until the food supply changes. Once losses exceed profits, people take notice.

The Cheapest of Thrills

It's now time for an economics lesson. Which addictive substance is the cheapest to procure, yet the most expensive burden to society? Nicotine used to be the cheapest. At its worst, lung cancer claimed 443,000 people a year, and cost health care $14 billion annually. But it also made society money, because the average smoker died at age sixty-four, before they started collecting from Social Security and Medicare. And through the 1970s, even with the removal from television airwaves, Big Tobacco still cornered the marketing market. The Marlboro Man, gorgeous ladies with their Virginia Slims, even doctors extolling the medical benefits of smoking. Cigarette taxes netted $25 billion for the government. How about alcohol? Each year, alcohol causes ten thousand deaths from drunk driving, twenty-five thousand deaths from cirrhosis and other diseases, and costs the medical system $100 billion annually. But it makes $5.6 billion per year for state and local governments in taxes.

Far and away, the most expensive burden is sugar. Because it wastes $1.8 trillion in health care spending, and kills slowly, thus reducing economic productivity. And of course it's the one that is subsidized by the federal government. Witness the tripling of global consumption in the last fifty years, and the doubling per capita in the United States. Amsterdam is the drug capital of the world. In 2013, Paul van der Velpen, chief health officer of the Netherlands, famously declared sugar to be "addictive and the most dangerous drug of the times."[26] Sugar is indisputably the easiest hedonic substance to procure, and despite its ubiquity, one that continues to fetch a higher price. And we pay for it twice. First we pay in federal subsidies (the price on anything that's not subsidized must therefore be raised to pay for the subsidy). Then we pay for the ER visits. We will see how all of society pays in Chapter 15.

But the public is catching on. In 2014 an NBC News/*Wall Street Journal* poll asked, which substance is the most dangerous to society? Forty-nine percent said cigarettes, 27 percent said alcohol, 8 percent said marijuana—all understandable responses. The surprise was that 15 percent said sugar was the most concerning substance. The word is getting out. Witness the spate of soda taxes being enacted all over the world, in an attempt to override the market and its marketing mayhem.

And just in time. The loose confederacy of Washington, Wall Street, Las Vegas, Silicon Valley, and Madison Avenue has brought us to this precipice. One step more, and the downward force will drag us all into an inescapable vortex, the proverbial *death spiral*.

15.

The Death Spiral

The "death spiral" is a doubles figure skating maneuver in which the larger skater (read: the man) pivots on one toe while holding the hand of the smaller skater (read: the woman), who whirls around him in a supine position as her head reaches closer and closer to the ice. Centripetal force pulls her back up. That's where America finds itself now, in a death spiral—a virtual health care vortex, flirting with oblivion, from which it seems impossible to escape. But there's no force trying to pull us back up; in fact, the centrifugal forces just push us closer to the brink. Obesity, diabetes, heart disease, stroke, cancer, and dementia are among a set of diseases known collectively as metabolic syndrome, and they threaten to bring our entire health care system to its knees. They are certainly devastating to the people suffering from them. Over 88 percent of Americans surveyed said they would rather be healthy than rich, yet only 37 percent believe that they will enjoy good health ten years from now.[1] Furthermore, 80 percent of all respondents over fifty already have at least one disease of metabolic syndrome. If they would rather be healthy than rich, they're doing a pretty lousy job. But they aren't the only ones suffering. Treating each of these conditions is expensive and all of society pays the price in taxes and higher insurance premiums. Ultimately metabolic syndrome will be the straw that breaks Medicare's back. Currently 9.3 percent of the adult American population is diabetic, and another 40 percent is prediabetic. But this is not just our problem. The same struggle is being fought around the world. Great Britain, Australia, Japan, Mexico, and South Korea are all experiencing the same bankrupting of health care, due to the same diseases. And Saudi Arabia, Kuwait, Qatar, UAE, and Malaysia are now at 80 percent obesity and 18 percent diabetes. Even with all their oil money, they can't support and underwrite this level of illness.

A Legal Ponzi Scheme

Where could these diseases be coming from? Everyone assumes this is the result of the obesity epidemic: too many calories in, too few out. Children and adults are getting fat, so they're getting sick. But here's the rub: while 80 percent of the obese population is metabolically ill, that means 20 percent are not. They are what we call "metabolically healthy obese," or MHO.[2] They will live a completely normal life, die at a normal age, and not cost the taxpayer a dime. They are not contributing to the death spiral. Conversely, 40 percent of the normal-weight population harbors these same diseases, but they're not obese.[3,4] If normal-weight people get them as well, how is this related to obesity? Rather, we now know that obesity is a *marker* rather than a *cause* of metabolic syndrome.

There's no question that more people, especially the poor, are getting the diseases of metabolic syndrome.[5] This is in part due to population growth: when there are more people, there are also more sick people. But

2015 saw an increase in age-adjusted mortality rates for all the diseases of metabolic syndrome;[6] heart disease, stroke, Alzheimer's disease, diabetes, and kidney disease. And it isn't just that more people are getting these diseases (incidence); technological and medical advances mean that people are able to live longer than ever before, meaning there are more people who have the disease at any given time (prevalence). Yes, we can keep these sick people alive longer, but a longer life isn't necessarily a better one, and certainly not a cheaper one.[7] The increasing incidence, prevalence, and severity of all these chronic metabolic diseases; the price of their treatments; and the fact that patients can simmer along debilitated for years is why health care is going broke. Stanford economist Raj Chetty demonstrated that personal income directly correlates with life span: the less money you have, the quicker you die.[8] Yet neither poverty nor minority status underlies the reason for the changes in U.S. mortality statistics over the 2005–2015 decade. We keep old sick people alive longer, which costs a lot of money. This expense should be offset by healthy young people paying in, yet we're losing them to drug overdoses,[9] addiction, and complications from metabolic syndrome (see Chapter 9).[10]

This is felt in all branches of government, but nowhere more than in social security. Social security is a legal Ponzi or pyramid scheme. Lots of young people at the bottom of the pyramid pay in, with the expectation that the money will be there once they make it to the top. The only difference between social security and Bernie Madoff is that the U.S. Treasury holds the money and, if you're lucky, pays it out. But social security is in crisis in every country. In order for social security to be healthy, you need lots of healthy young people at the bottom paying in, with the fewest old sick people taking it out at the top.

Social security in the U.S. started to falter in the late 1990s. Before that, our government had the ideal paradigm. We had a nation of smokers; even though we knew that U.S. surgeon general Luther Terry's 1964 report categorically demonstrated that smoking kills,[11] the U.S. government did little to curtail smoking for the next thirty years. Why? Because the actuaries determined that smoking killed you at a mean age of sixty-four—one year before you started to collect social security. And the insurers were delighted, because you could only spend so much money on lung cancer, which would kill you in an average of six months. So you had healthy young people paying in and then dropping dead just before they started to take out. The ultimate Ponzi scheme.

But, instead, what if the young people are sick? What if the young people are on disability due to chronic disease? Now we're talking about diseases that don't kill you quick, like lung cancer. We're talking diabetes, fatty liver disease, heart failure, kidney failure—diseases that take twenty years of misery and money before you succumb. Worse yet, what if social security is paying out to all these debilitated young people in benefits? When the base of the pyramid crumbles, the whole structure collapses. And that is what is happening right now, all over the world. My good friend Juan Lozano Tovar, currently head of social security for the government of Mexico, convened a symposium on the future of international social security at MIT in May 2015. At this symposium we discussed economics, pensions, aging, and genetics.[12] The elephant in the room was the epidemic of chronic disease in young people. I see ten-year-olds with type 2 diabetes every day; they'll likely be on long-term disability and never hold a job. When the money goes out instead of coming in, everything else is rearranging the deck chairs on the *Titanic*.

Eating Away at the Base of the Pyramid

One big portion of the problem is that we have focused all of our efforts on health care rather than health, on treatment instead of prevention. We plow billions of dollars into drugs and surgeries and nutraceuticals, some of which have in fact reduced death rates, but none of which has reduced the actual rates of disease. In the good old days, we got sick and we died. (Is that good? It is if you want to keep a health care system afloat.) Today, fewer people are dying from heart disease or diabetes, because we have treatments to keep them alive.[13] Now they hang around, with compromised worker productivity, detracting from the positive side of the economic ledger. Yet they rack up higher medical costs, increasing the negative side of the financial ledger as well. In fact, all of our medical technologies have allowed us to tread water, but they haven't rescued us from the pool. We've just delayed the drowning. We are literally in the throes of the death spiral, with a vortex so strong, we can't escape the forces dragging us downward.

Why is this happening? What is the cause of this death spiral? Either we're so unhappy that we are killing ourselves (see Chapter 11), or we're under the malicious spell of some external force that is powering this tornado. Are our opioid crisis and our metabolic syndrome crisis related? What is the connection between happiness, health, health care, and life span?

Happily Ever After

There's no question that innate happiness predicts a longer life. A group of American nuns were asked to write an autobiography in their twenties, and their writings were analyzed for content on well-being and positive affect. Those who exhibited contentment lived to an older age than those who did not.[14] But was that because those who were happy didn't find the need to engage in more problematic behavior, e.g., smoking or alcohol (neither of which is forbidden by the Church)? OK, but those are nuns. A more recent and complete assessment of different populations internationally[15] suggests that high subjective well-being and life satisfaction predicts longevity irrespective of economic status, although it doesn't necessarily extend the lives of people who are afflicted with chronic disease.

Conversely, poor health clearly causes unhappiness. And poor health is a primary predictor of mortality. But does unhappiness cause mortality directly? In the case of suicide, unequivocally yes. And those who are stressed, unhappy, and yearning for contentment often seek a consolation prize in the form of reward, thus breeding a host of lifestyle factors associated with skewed dopamine, such as alcohol, tobacco, and street drugs. So which kills you—the behaviors or the unhappiness? Three recent studies, two from the UK and one from the U.S., start to answer this question.

First, a British longitudinal study of men and women ages fifty and older[16] modeled the role that happiness plays in predicting a long life. Happiness was assessed using a four-point questionnaire (I enjoy what I do; I enjoy the company of others; I look back with a sense of happiness; I feel full of energy). Of note is that this study took into account both depression and the diseases of metabolic syndrome (e.g., heart disease, stroke, diabetes, cancer, impaired mobility, chronic lung disease) in the model, but did not ask about the behaviors that led to these diseases (sedentary activity, bad food, smoking, alcohol, drugs, sleep debt). The results showed that of all those who had specific diseases, you were more likely to survive them if you were happy by a factor of 25 percent.

However, the conclusion reached in the next study was somewhat different. In the UK Million Women Study,[17] post-menopausal women at a median age of fifty-nine years were asked, "How happy are you?" About 40 percent said they were happy most of the time, 44 percent said some of the time, and 17 percent said none of the time. They asked these women to self-rate their own health. The investigators followed this group for ten years. Death rate was definitely associated with poor health assessment at baseline, and poor health was associated with unhappiness. But after adjustment for self-rated health, reward-driven behaviors (smoking and

hedonic eating), and treatments with medicines for reward-related diseases (e.g., hypertension, diabetes, and also depression and anxiety), it
turned out that these women's innate unhappiness played no role in their
death rate from heart disease, cancer, or any other disease.

The third U.S.-based study adds yet another piece to the puzzle. A
group of adults had fMRIs performed at baseline and then three to four
years later, during which time 9 percent of the subjects suffered a heart
attack.[18] What was different about their brains that could be a predictor?
Those who demonstrated the highest activities in the amygdala (the fear
center) were the most likely to experience a cardiac event. The amygdala
dampens function of your PFC (see Chapter 4), putting you at risk for
more self-destructive behaviors.

What are we to make of these three seemingly different yet related
studies? They tell us that unhappiness itself isn't the killer. Rather, it is
likely that our fear and anxiety stoke our unhappiness, which drives us
toward unhealthy behaviors (many of which are dopamine driven), and it
is these behaviors that increase morbidity, disability, and eventually mortality. These studies don't answer directionality—which came first, the
fear or the unhappiness? Ultimately, it doesn't matter: behaviors continue
to increase the force and the expense of the death spiral. It's all about how
we seek pleasure in the absence of contentment, driving both addiction
and depression, and where and how we have run off the rails. It's the
cheap thrills that kill. And the cheap thrills are everywhere. But there's a
silver lining here, because all of these cheap thrills are man-made. And
they could be un-made—if we choose to.

The Tragedy of the Commons

Health care and social security are finite resources. There's only so
much money in the pipeline. In the past, we've just upped the dollars

devoted to health care. In 1965, health care amounted to 5 percent of gross domestic product. In 2014, health care took up 17.5 percent of GDP. And by 2022 it is projected to be 19.9 percent. But now we can't print money fast enough to afford it; the money has dried up. When such finite monetary resources are used indiscriminately by everyone, we suffer from the paradigm known as the "Tragedy of the Commons." The principle: you have a big field where all the farmers graze their cattle. When there are few cattle, there's no problem. But when farmers buy more cattle and let them graze to their hearts' content, soon there's no grass and no cattle. This is a well-documented principle by which a limited resource that is supposedly available for all is utilized by all and then disappears for all. This is what is happening to social security and health care. And ostensibly, this was the basis for Obamacare.

Dr. Ezekiel Emanuel, President Obama's health care advisor and architect of Obamacare, published an article in October 2014 in the *Atlantic* titled "Why I Want to Die at 75."[19] In it he argues that health care has not extended the time of living but rather has extended the time of dying. While we have increased the length of life, we have not contributed to improving either happiness or productivity.

The data would argue that our paradigm of life (or death) extension seems to have reached its zenith. Newer research argues that this is the first generation to live shorter lives than its parents.[20] In Chapter 11 we noted that life span has started to decline. While the percentages might not seem like much, it's the slippery slope. Expect further life span reductions in the future, despite spending over one-fifth of our GDP on health care.

Worse yet, Obamacare was predicated on a false premise—that the healthy people would pay for the sick people, and that access to a doctor would keep them out of the ER. The idea was that we could add 32 million sick people to the insurance rolls, and we would pay for this by providing "preventative services"—that is, by having access to a doctor, and by treating symptoms with medicines (e.g., treating high cholesterol with statins, high blood pressure with angiotensin converting enzyme [ACE] inhibitors, diabetes with oral hypoglycemic agents), you would stay out of the emergency room, where the costs are fifty times higher. Sounds good, but there's a catch. There are no services to prevent chronic metabolic disease that can be provided by the health care system. There's only treatment. We doctors can keep these people alive, but we can't stop people from getting heart disease or diabetes or fatty liver disease or kidney disease with pills, just like we can't stop people from getting obese. And these treatments cost health care big bucks. Just look at how the rates of these diseases have continued to climb over the past twenty years, despite our full knowledge of the obesity epidemic. And once you've had your heart attack or stroke or once you're in chronic kidney failure, you're nothing but a liability. You're just another head of cattle grazing on the commons. Why do you think three of the biggest health insurers (United, Aetna, Humana) opted out of Obamacare: at a 15 percent profit cap, they couldn't keep up with all the sick people with diabetes. Yet we could all save money on chronic metabolic diseases if we could actually prevent them.

Health Care Is Sick Care

Health care has been sucking at the teat of the federal government for decades. Everyone wants their share. Doctors, lawyers, and insurers were natural-born enemies. After all, we doctors wasted the insurance company's profit on health care and its delivery. We demanded all modes of exorbitant therapy for our patients, including treatments that might prolong a patient's life for a few months. Never was this truer than in the good old heydays of health insurance. Big Medicine relied on Big Treatment to generate Big Bucks. Treating chronic diseases, especially those with no cure, is expensive, and the markup is huge. The government negotiated with the American Medical Association (AMA) to provide fee-for-service. Procedures became the cash cow, and the premiums reflected it. The doctors drove the bus, the insurance company charged higher rates, the employers anted up, and the patients came to expect the magic of modern medicine. No talk of prevention. The results of prevention don't happen within one political cycle. Prevention requires a cultural shift, and that is political suicide, especially when there's money to be made. Prevention puts responsibility not just on the individual but on society as a whole, including government. Prevention may sound good on a bumper sticker, but it doesn't make any money for doctors, for hospitals, for insurers, or for politicians for that matter. Prevention is a zero-sum game.

The costs associated with procedure-driven health care made doctors the bad guys through the last half of the twentieth century. We doctors lost our credibility with Congress. This was the impetus for expansion of HMOs: to try to control doctors and, in turn, control costs. Except 43 percent of health care was spent on administration by the insurance companies. So who really wasted the money? But hey, just raise the premiums. Keep those profits rolling in. And the lawyers were happy to tap into all that largesse. Malpractice suits generated awards that topped $69 million a pop. At one point, almost 60 percent of all doctors had been sued for malpractice at least one time in their careers. All paid out of the insurers' profits. And so it was.

But the Tragedy of the Commons means that the party's over and the grazing land has gone fallow. Academic medicine, Medicaid, Medicare—they're all on life support. And the bankruptcies are not due to physicians' salaries, which have decreased relative to inflation, or in-patient costs, nursing costs, or infrastructure expenditures. It's due to chronic metabolic disease. These diseases aren't killing us outright; instead they're sucking us dry. And if you think that other people getting sick is their problem and not yours, chew on this: 65 percent of all health care expenditures are paid out of government dollars. That means your taxes.[21] You're paying twice—more for your own insurance premium, and more for everyone else too.

One question around which the 2016 election revolved is whether Obamacare worked. Well, as of 2016, 20 million people who were previously denied or could not afford it were able to obtain health insurance. Insurance company profits were capped at 15 percent. Any extra money that the insurance company took in had to be returned to its subscribers. So that should reduce corporate profits, right? Not necessarily.[22] The large insurance plans have cut administrative costs to make the 85 percent minimum loss ration (MLR), but they hiked both premiums and deductibles, so that it became harder to use the health care you got. This resulted in the big insurers making even more money and doling out bigger executive bonuses.[23] Those three major insurers abandoned the state exchanges altogether, thus squeezing the rest. Because with 15 percent profit, they can't make money using the casino model anymore. The best way to make money now is to *save it*; that is, nobody get sick so there's no payout (or make the premium so high that nobody uses it). And Trumpcare will continue to ignore the elephant in the room: more of the populace is sick, and getting sicker. And they won't get better until there's a substantive change in the food supply. Trumpcare's answer? Deny 24 million people health insurance.

There is one good thing that did result from Obamacare that will likely remain, perhaps an unintended consequence. Before Obamacare, insurance was based on the casino model. It was pay-to-play, and the insurance company, like the casino, set the payout. Under the casino model, the insurance company *wanted you to get sick*. More people getting sick was their excuse to raise the premiums and make more money. There was no impetus for prevention because there was no money in prevention. And decades ago, Big Business wasn't necessarily happy when people got sick, but it wasn't crying either as CEOs could replace older employees, who garnered high salaries and pensions, with younger ones to whom they could pay lower salaries and who had never heard of pensions. But as the premiums grew, Big Business realized that insurance costs were cutting into their economic productivity, because they are spending $2,751 per annum per employee for obesity-related diseases, whether the employee is obese or not.[24] Nonetheless, because of the cap, for the first time in history, insurance companies actually *want you to be healthy*. So they are all paying for preventative care now. Now they want their payout to go down, and a healthy workforce is the only way to reduce those costs.

Stayin' Alive

But what if we had healthy ninety-year-olds instead of sick seventy-five-year-olds? My wife's grandmother lived on her own farm in rural Minnesota till the ripe old age of 101, growing her own food, gardening, limiting her TV, and not seeing a doctor. Her family's single regret in her life was that when she hit 100, the *Today Show*'s Willard Scott didn't mention her. What if we all consumed minimal health care resources in our old age because we were the epitome of health? What if we prevented the national exposure that drives economic, social, and medical devolution all at once? And what if we could be happy at the same time? In Zeke Emanuel's *Atlantic* article, he never even addressed the biggest issue in this entire debacle: diet. Could we turn this around? Could we save lives *and* money? We could, if we would just reduce the most ubiquitous dietary item that is driving it upward: the cheapest of thrills—sugar.

While there are likely many environmental factors involved in metabolic syndrome, the one we have causation for is sugar.[25] Disability-adjusted life years, or DALYs, are a way of determining the health and economic burden of a particular exposure or behavior. One study[26] looked at the effects of sugar-sweetened beverages on DALYs worldwide. What was interesting about this analysis was the breakdown by age group. Despite having the highest general disease rates, the over-sixty-fives were virtually untouched by sugar-sweetened beverages. Conversely, it was the twenty- to forty-four-year-old age group that showed massive increases in DALYs due to sugar-sweetened beverages. In other words, drinking sodas, juices, and sports drinks doesn't just hurt you when you're older. It hurts you now, when you're supposed to be at top earning potential, when you're supposed to be paying into the system.

In America, it's no better. The Supplemental Nutrition Assistance Program (SNAP), also known as food stamps, is a $75 billion program that covers 15 percent of adult Americans and 33 percent of children. What do these people buy? Sugared beverages are number two, and some other form of sugar-containing food are numbers four, five, ten, eleven, and twelve; amounting to 27 percent of all expenditures.[27] Why should we care what they buy with our taxpayer money? Because people who get their food on the SNAP program are 50 percent more likely to die of heart disease or diabetes than SNAP-eligible people who don't participate, and three times more likely than those who are SNAP ineligible.[28]

My UCSF global health team looked at what would happen to the death spiral if we just cut back a little on our cheapest thrill, sugar.[29] From our research on metabolic syndrome, we identified the development of liver fat as the single most important risk factor for developing the various diseases of metabolic syndrome. We then modeled how disease prevalence would change, and how much money could be recouped, if the United States engaged in a campaign to reduce our added-sugar consumption, first by 20 percent (the amount that is targeted by sugar-sweetened beverage taxes, such as the one in Philadelphia), and then by 50 percent (which would bring us to the 10 percent added-sugar limit recommended by the new USDA dietary guidelines). The model was carried out going forward for twenty years, from 2015 to 2035. The results were astonishing. For instance, at a 20 percent reduction in added-sugar consumption, cases of heart disease and diabetes in the U.S. would decrease by 0.1 percent and 0.2 percent, with health care savings of $10 billion, while a more restrictive 50 percent reduction in added sugar consumption would reduce heart disease and diabetes by 0.3 percent and 1.2 percent, with combined health care savings of $32 billion per year. Over a twenty-year year period, we're talking about a half trillion dollars saved. Such a move would be a boon for both health and health care. And what is driving increased sugar consumption? It isn't a slam-dunk, but in most people dopamine plays an outsized role.

Redrawing the Lines of Engagement

Politics makes strange bedfellows. Natural-born enemies aren't such enemies anymore. It used to be the doctors against the lawyers, the doctors against the insurers. But now we're all on the same side for the first time. It's all the people who stand to lose—the doctors, the lawyers, the insurers, and those big businesses that have to pay the health care tab—against those who want to maintain the status quo—the food industry, the drug industry, the White House, and Congress. All of a sudden we have some very rich partners with a lot of clout. Now we just have to figure out how to harness it.

Health care is just the tip of the iceberg. The economic and social devolution of the last forty years is unsustainable and is catching up to us now. Why is this happening? What are we doing wrong? The short answer is that America has lost its way. The problem runs deep, like a sewer, and it stinks. The cause and cure of what ails us lies not in the art of medicine or the politics of health care but rather in the science of happiness. And our wanton desire—for anything and everything—has betrayed us.

The death spiral derives its strength from our collective unhappiness, which fuels its centrifugal force. But you *can* swim yourself to safety. Part V will show you show.

PART V

Out of Our Minds– In Search of the Four Cs

16.

Connect
(Religion, Social Support,
Conversation)

P hilosopher Eric Hoffer was quoted: "The search for happiness is one of the chief sources of unhappiness." I couldn't agree more— the destination is not the journey. Pixar chief John Lasseter said, "The journey is the reward." The problem is, if you keep making a series of wrong turns, you could end up horrendously lost. Yikes. Is it any wonder that happiness seems so elusive?

First you have to recognize what happiness really is. In the first fifteen chapters of this book, I hope I have made a cogent case that: (1) reward is not contentment, and pleasure is not happiness; (2) reward is dopamine, and contentment is serotonin; (3) chronic excess reward interferes with contentment; (4) business has conflated pleasure with happiness consciously and with clear-cut intent, specifically to get you to buy its junk or engage in hedonic behaviors favorable to industry; (5) government has passed legislation to make it easier to buy that junk or make easier access to engage in those behaviors to drive profit and GDP, and the Supreme Court has justified and supported these practices; and (6) buying that junk or engaging in those behaviors long-term and without thought can leave you and society fat, sick, stupid, broke, addicted, depressed, and most decidedly unhappy.

Consumption is related to the health of an economy, but clearly not the index of health of a society. It's pie-in-the-sky to think that the U.S. political system (or any other, save for Bhutan's) is going to decide that individual or collective happiness outstrips GDP as the prime indicator of progress. Business certainly contributes to the pleasure side of GDP, and government has codified it. Except that pursuit of GDP loses more money for society than it makes for individuals, and so the death spiral continues. The only way to slow or stop the vortex is to increase individual contentment and, as a result, societal happiness. But for happiness, we're on our own to pursue it—or not. And therein lies the rub: you *have* to pursue it. How? Where to turn?

The next four chapters will provide the GPS for happiness, based firmly in science, so you can't get lost. While none of these modalities will in and of themselves fix the systemic problems of our corporate consumption society, they do have the power to help you ramp up your serotonin, tamp down your dopamine and cortisol, and reclaim your happiness and, in turn, your life. Here now, I delineate the Four Cs of Contentment: *Connect, Contribute, Cope,* and *Cook.* The rationale for each is bolstered by their documented neuroscientific effects on our three limbic system pathways—the reward pathway, the contentment pathway, and the stress-fear-memory pathway (see Chapter 2). When used properly, each has been proven to be clinically effective on its own and even more so together. You can perform any or all of these—without a prescription, without a personal trainer, without cost, and at home.[1] But there are two caveats to their efficacy:

1. None of these four modalities is passive: you have to *perform* them for any to work. The pursuit of happiness is *active.* As you will see, in some cases *pursuit* means *actively doing nothing.*

2. Each one of these Cs has been co-opted by various industry actors
   in an attempt to subvert your efforts. They want you to believe that
   they've cornered the market on happiness and that you'll want
   what they're selling. In fact, the happiness industry has been born
   out of the anxiety fomented by our current stress-happy environment. In the case of medicine, the wellness industry was born out
   of the fact that pharmaceuticals can treat but not prevent disease.
   Therefore it is incumbent on me as we progress through these next
   four chapters that I point out these wrong turns down the road to
   happiness so you don't become completely discombobulated.

Perception Is Truth

Recall the hallucinogen studies from Chapter 8; the psychedelic experience is mediated by the serotonin-2a receptor, but the feelings of contentment appear to be due to cross-reactivity of certain hallucinogens, such as
LSD, with the -1a receptor. The genetic studies, depression studies, and
pharmacology studies (see Chapter 7) all say the same thing: for contentment, or serenity, or *eudemonia*, or subjective well-being—whatever you
want to call it—serotonin has to bind to its -1a receptor. Clearly, complex
mood disorders are influenced by serotonin levels, with the -1a receptor
playing a major role. One group of investigators evaluated serotonin-1a
receptor density across monozygotic (identical) and dizygotic (fraternal)
twins, and determined that serotonin effects are not determined by genes
alone; they can be responsive to the environment.[2] That means you have
at least some nominal control over your perception of contentment. For
those of you with pets, no doubt you'll appreciate this metaphor. If you
want a happy house, first make sure your cat Serotonin is purring. And
then, while you're at it, keep that darn yappy dog Dopamine from

constant overstimulation, or else he's going to pee on the rug, ruin your party, smell for a while, and possibly discolor your interior permanently.

An Act of Faith

The hallucinogen studies provide one window on how to get there: both the mystical experience and contentment are dependent on different signaling pathways of the serotonin molecule. Well, what can offer up mystical experiences without drugs? One place is religion. Interestingly, Karl Marx called religion "the opiate of the people," placing it squarely in the reward/pleasure pathway. But at its best, religion impacts your serotonin and can bring contentment. As you'll learn in this section, when it activates dopamine, not so much . . .

People turn to religion as a gateway to happiness for at least sixteen different reasons, including acceptance, power, curiosity, order, idealism, and the concept of "self-transcendence," or belonging to something bigger than yourself.[3] It is no surprise that many religions have attempted to harness that social connection to benefit the greater good, another potential pathway to contentment (see Chapter 17). The concepts of the Jewish *tikkun olam* (healing the world) and Christianity's "faith without good deeds is dead" (James 2:14–26) promote collective well-being with "trickle-down" personal well-being (the opposite of "trickle-down economics"). It is in the giving to one's children, one's family, and to others that happiness can be realized. Eastern religions such as Buddhism and the Baha'i faith similarly endorse the concept that the true path to happiness is determined by what you do for others. Without doubt, one major appeal of organized religion is its basis in community, sharing a collective belief/purpose with like-minded people, attending services, and/or simply knowing that a group supports you. Another potential reason for religion's appeal, as Freud postulated, is to ward off anxiety about death by affirming an afterlife. Reducing anxiety (stress and cortisol) may increase serotonin-1a receptors and yield serotonergic benefits in the form of contentment (see Chapter 10).

Many people pursue their paths to individual happiness through religion. Yet, over the past two decades, more people (not just Americans) have either been taking paths away from religion[4, 5] or changing their path from one religion to another.[6] Is secularism winning? Is religion not working? Maybe it's just *your* religion that's got problems . . . In fact, the psychological literature suggests that religious people do tend to be happier. But as we've already learned, it depends on how you define happiness. The U.K.'s Office of National Statistics computed the Kahneman-Deaton Life Satisfaction Index of religious people, who scored only slightly higher than non-believers.[7] One study looked carefully at these data, and there was an interaction between "strong religious identity" and "building social networks within their community";[8] for those with weak religious identity, the social network was meaningless. Other studies looked at the relationship between religion and scales of subjective well-being, which tells a slightly different story. The Gallup Organization conducted a poll of 676,000 "religious" people to determine who had the highest subjective well-being. Answer: Jews and Mormons.[9] Really? A good proportion of American Jews are secular rather than religious and they tend to kvetch a lot, and Mormons, well . . . they're Mormons. They're *born* smiling. But the data suggests at least three areas of overlap. Both are pro-social and family-focused, and emphasize purpose in life contributing to the greater good.[10] Interestingly, the tenets of twelve steps–based addiction programs are pretty similar. An international analysis found a relationship between religiosity and subjective well-being, which demonstrated an interaction between social support and societal circumstances. That is, in impoverished countries, religiosity-predicted social support that predicted subjective well-being.[11] Social interaction seems to be the crucial factor in both indices generating contentment.

Of course, religion also serves a very important practical goal, as this book argues, by making sure reward and contentment remain mutually exclusive (e.g., Onan spilled his seed and God slew him [Genesis 38:9]; the Israelites worshipping the golden calf [Exodus 32:5]; Christian suppression of pleasure with the promise of heaven; Sharia law forbidding alcohol, tobacco, porn, and gambling, and Mormons add coffee to that list). Also, by invoking an afterlife, many religions emphasize the "long" game, rather than the "short" one—although very few practitioners of any religion adhere faithfully.

So do our three limbic pathways help to explain the effects of well-being in religious people? Neuroscientist Sam Harris, deemed one of the "four horsemen of modern atheism," scanned the brains of fifteen devout Christians and fifteen atheists while asking them true/false questions about their beliefs and judgments on such things as the Immaculate Conception and the Resurrection.[12] Unrelated to belief status, whenever the subject felt the statement was true for them, the prefrontal cortex (PFC) lit up, indicating cognitive thought and endorsement. Other areas of the brain lit up, though inconsistently. This finding argues that belief is a cognitive, or thinking, process rather than a visceral or subconscious one, and that this brain activity is independent of belief status.

How about serotonin and dopamine? If serotonin makes you content, does it also make you religious? One group performed PET scanning in fifteen healthy subjects using a radiolabel (a radioactive compound that bound like serotonin to the -1a receptor so they could quantify the binding) and found a relationship between "self-transcendence" and "spiritual acceptance" in the DRN (the home of the serotonin neurons), the hippocampus (the home of memory), and the cortex (where thoughts are processed).[13] Indeed, certain genotypes of the serotonin-1a receptor are related to these qualities of religiosity.[14] So there appears to be some evidence for a role of serotonin influencing religious beliefs. Of course, these are small sample sizes, but the directionality is consistent and therefore worth examining.

Conversely, dopamine dysregulation is a hallmark of untreated schizophrenia, a form of psychosis; drugs that antagonize the dopamine receptor (e.g., risperidone) are potent antipsychotics. Schizophrenic patients often attest to hyperreligiosity, as their auditory hallucinations can involve hearing God or angels talking to them. In a small study, hyperreligiosity was seen more commonly in schizophrenics than in other causes of psychosis.[15] Parkinson's disease patients, after they are treated with L-DOPA (the precursor to dopamine), have often been noted to exhibit increased religious fervor. It has been proposed that dopamine might be the trigger that takes a person from believer to zealot.[16] Of course, the relationship between serotonin, dopamine, and religion is limited to correlation, not causation—speculation, theories waiting to be proven—but it's clear that the role of biochemistry in the mediation of the religious experience will remain an important research question.

The one tidbit we can take away from this scientific exploration of religion is that it's not the incantations, it's not the incense, it's not the genuflecting—it's the social engagement or emotional bonding that correlates with contentment. When you are a part of something larger than yourself—whether united by religion, or tribal origin or heritage, or a worldview, or a hobby, or a common goal—you feel a greater sense of contentment.

On the Same Wavelength

We humans are engineered to develop social support—that is, emotional bonding in the form of interpersonal relationships,[17] starting with mother-to-baby and working forward over the decades. Social support has a strong evidence base for benefit for the individual and for society. Low social support is linked to progression of numerous illnesses and early death.[18] Social contact activates the PFC—which may thereby tamp down the amygdala, thus reducing stress and cortisol—and parts of the reward pathway are linked to various forms of caregiving (like the mother-child bond), which can increase endogenous opioid peptides (EOPs) that further dampen stress hormones.[19] There's even an early line of research that suggests that one's degree of interpersonal connectivity predicts im-proved cognitive functioning at different ages (including the elderly), and that just ten minutes of talking to another human being per day can re-duce your risk for dementia.[20]

People find contentment in being part of a community, which evinces social relationships. Social support correlates with positive emotions, greater reward activation, and increases in serotonin.[21] Conversely, low social support correlates with negative emotions such as hostility,[22] with less reward for social stimuli[23] as well as with low serotonin.

Have a Heart

Performing acts of compassion, like visiting a sick friend, provides a pow-erful sense of connection and is a prime promoter of emotional well-being and contentment. The effects of compassion even register in children's brains. One study looked at brain wave patterns in six- to ten-year-old children. When they were experiencing contentment, they dem-onstrated activation of the left PFC, and when they were experiencing empathy, the right PFC was more active.[24] Why does empathy travel with contentment? This remains a hotly debated topic. One theory posits that each of us possesses a network of "mirror neurons," a widely dispersed class of brain cells that work sort of like a neural Wi-Fi. Do an experiment: Call someone on the telephone (not FaceTime) and, somewhere in the conversation, state that you are waving hello. Then ask what they are doing. Chances are they're waving back. Mirror neurons take in visual, auditory, and tactile information, track the emotional flow, turn this sensed state into emotion, and then transmit it into our own brain to mimic these same emotions. Presumably, visiting a sick person will cheer them up; this will cause the visitor's mirror neurons to capture the sick person's joy and activate similar positive feelings in the brain.[25] The same is true about serving a meal in a soup kitchen. By improving the lives of others, you improve your own.

Studies have since attempted to validate this concept by identifying a phenomenon called "interpersonal synchrony," in which one person's actions within a relationship alter the experience and the emotion of the other: an empathic connection. Several areas of the brain are activated, but none more so than our old friend Jiminy, our PFC,[26] presumably telling the rest of the brain to relax, not be afraid, and to allow the good feelings to flow. This concept of interpersonal synchrony was put to the test by evaluating the responses of people with and without autism spectrum disorder. When the leader input information into a computer for the follower to obey, normal people, as expected, demonstrated synchronous behavior, empathy, and fMRI activation of the PFC. Furthermore, the degree of synchrony correlated with the degree of empathy. However, subjects with autism were unable to demonstrate this synchronous behavior or generate empathy, and their PFCs did not light up,[27] suggesting that autism may be a problem of PFC functioning. Thus, the PFC appears to coordinate the emotional response to interpersonal connection. Such coordination of emotions, cardiovascular reactions, or brain states between two people has been studied in mothers with their infants, marital partners arguing, and even among other people in conflict.[28]

It would appear that with interpersonal synchrony, the biology of one person can alter the biology of the other. As an example, the Framingham

Heart Study started in 1948, and continues to examine the natural history and predictors of chronic disease in the U.S. population. One of the most jarring findings in recent memory was the demonstration that obesity could be "transmitted" within social groups. Usually when we think of disease spreading from one person to another, we think of some kind of infection. But in this cohort, obesity was related to whom you hung out with: your friends determined your weight, possibly because you're eating the same types of food with them, good or bad.[29] These same investigators have also shown that happiness could spread throughout social networks in a similar fashion.[30] If you have a happy friend, spouse, sibling, or neighbor who lives within one mile of you, you have a 25 percent chance of being happier. Furthermore, the effect diminished as you moved farther away. What does this mean? It means that it's the *social* part of the social network that allows for the transmission of happiness. However, there is a big caveat to this analysis: the data for this study was collected prior to 2003. Facebook was not founded until 2004.

Fear of Rejection

So what happens when your social connections rebuff you? What happens when your boyfriend, your debate team, your coach, or your social network abandons you? Is there biochemistry behind emotional rejection? Volunteers underwent fMRI of the brain under two conditions, in random order. One time they had a very hot compress applied to their upper arm. The other time they were shown a picture of their ex. As expected, the hot compress activated the parietal lobe on the opposite side of the body where the pain is registered (because pain fibers cross sides in the brain stem). But otherwise the activation of virtually all parts of the brain completely overlapped in both conditions. Various portions of the limbic system (the emotional part of the brain) were activated, due to either physical pain or social rejection.[31] One theory is that the brain's pain centers may have taken on a hypersensitivity to social exclusion, because until the nineteenth century being an outcast was tantamount to a death sentence. These studies give new meaning to having an "achy, breaky heart."

The Social Network?

Clearly, developing and maintaining interpersonal connections is good, while severing them is bad. No surprise there. But in the twenty-first century, what constitutes an interpersonal connection? In each of these examples, at least two people are involved—*in person*. What if they are online instead? If you're not interacting face-to-face and in real time, is it still interpersonal? Does it still qualify as a social connection? Do you still get the benefits? Social media now dominate the landscape of human interaction. They are enough to spark a revolution; witness the Arab Spring. But are they enough to spark the same sense of affiliation, incur the same level of social support, and generate the same level of contentment? Are social networks really *social*? How many times have you preferred texting to talking? Are you getting the same return on your investment?

Here's where you've been sandbagged by the technology industry. The social media companies say they provide connectivity like never before, and at the speed of the internet. Facebook is now used by a total of 1.7 billion people—that's 25 percent of the world's population—and 1.1 billion log in at least once a day to engage in socializing, entertainment, self-status seeking, and information gathering.[32] Mark Zuckerberg says, "Our mission is to make the world more open and connected. We do this by giving people the power to share whatever they want and be connected to whoever they want, no matter where they are." Connected, sure. But interpersonal? Can you have an interpersonal connection with Anonymous? Do emojis convey empathy? And if the connection is not interpersonal, can you generate mirror neurons, synchrony, PFC activation, empathy, contentment, or serotonin? Are we all just, as MIT's media researcher Sherry Turkle surmises, "alone together"? Or are just some of us alone? How many Facebook friends do you have? You may well know the names, ages, and sports preferences of your middle school acquaintance's children, but could you actually carry on a conversation with them? Would you meet them for coffee, and what would you say to them, to their face? Do these connections transfer over IRL (In Real Life)?

We finally have data to start to answer some of these questions. By following adolescent boys longitudinally over time, scientists have worked backward to see what kinds of baseline behaviors contributed to those who developed internet addiction in midlife (mean age forty-three).[33] OK, so let's envision our forty-something guy addicted to the internet. What's he doing? (1) Looking at porn, (2) gambling, (3) playing online video games, (4) uploading snarky content to 4chan or Reddit. He's not obsessively surfing the Web for cat videos. These investigators could chart a path from parent-child conflict to problem behaviors like alcohol and drug use, as well as dysphoria and, in some cases, depression. Indeed, the internet could just be the legal behavior of choice for those with undiagnosed depression, because it provides an extra dopamine boost. The fact that those who would succumb to internet addiction could be predicted by personality traits in earlier life would suggest that internet addiction is the end result of other problem behaviors or psychiatric disturbances, and possibly even another example of "addiction transfer," rather than that the internet itself was the cause of these behaviors (see Chapter 5).

The question is, can social media fulfill your need for interpersonal connection? Studies of large social networks now suggest that reading postings of others' emotions can affect your own (also implying the presence of mirror neurons). When Facebook's news feed was manipulated by the company to post more positive emotional content, the response consisted of more positive comments, and vice versa.[34] Similarly, when there was more rainfall in any given locale, the postings out of that locale were more negative and the comments in response tended to be more negative.[35] Of course, these types of studies don't account for the possibility that people who post on Facebook are already self-selecting and may have different emotional needs than other people, and they don't tell us about the quality or depth of these emotions or for how long their emotions are impacted.

Facebook does appear to be an outlet for emotion, particularly in depression. For instance, the level of depressive symptoms obtained on a questionnaire in a group of "normal" subjects were directly correlated with the volume of negative postings on Facebook,[36] presumably as a cry for help[37] (of course, whose definition of "normal" are they using?). But does that help come? Yes and no. One study evaluated twenty-one patients (mostly women) with major depressive disorder (MDD) compared to twenty-one "normal" subjects.[38] Those with MDD were more likely to post deprecating (toward themselves and others) information on Facebook, and received social media support for those posts. However, the perception of the MDD patients was that they received less social support than they actually did: there was no relation between MDD, positive postings, and social support. So did they get emotional support or not? Objectively yes, but subjectively no. Which is more important? Without the face-to-face interaction, a one- or two-sentence comment or sad-face emoji doesn't help matters any. Are these MDD patients asking too much of Facebook? Is that Facebook's fault?

Well, maybe. Many people use Facebook for posting lovely smiling photos of their families, friends, and vacations. But could the use of Facebook actively undermine your feelings of subjective well-being? And in the extreme, could it even make you addicted or depressed or both? A two-week time-lag analysis[39] suggests that the more people use Facebook, the less subjective well-being they experience. This study was also able to show that direct interpersonal interaction during that same two-week period was able to improve their affective (subjective) well-being—yet the Facebook effect on negative well-being still persisted. In other words, being "social" on Facebook meant being less social everywhere else. So what is it about Facebook that makes people unhappy? A one-week study in which people's Facebook usage was monitored demonstrated that only passive Facebook usage (reading others' posts, and not adding your own) predicted declines in subjective well-being. However, the converse was not true: posting didn't make depressed people feel better.[40] Many in the psychiatric field have posited that Facebook actually makes us less content. Think about it—in Japan, a photo service called Family Romance sends fake friends to your house to take pictures for posting to unnerve your ex after you've broken up. Who's supposed to feel better after that? Drilling down, investigators have shown that feelings of envy from reading about other people's positive experiences worsened the moods of the readers. We usually see the very best of our peers, and so we constantly compare ourselves to an unrealistic and untrue ideal. Aside from negatively impacting mood,[41] what social media really does is shut down our PFC (our Jiminy) so we can't ratchet our emotional level down (see Chapter 4). How many negative comments are there when someone posts about *their* opinion, especially during the 2016 election cycle? The reaction is akin to ramming someone's car in the parking lot (Towanda!).

Finally, can Facebook be addictive? One theory argues that some individuals prefer to communicate online and demonstrate deficient self-regulation of internet use. They tend to engage in online social communication as a means of escaping loneliness or anxiety, a process that generates reward, which reinforces online use.[42] A meta-analysis of Facebook usage identified that excessive use of social media is the underlying process for gratification of need, which can devolve into Facebook addiction.[43] Is it any wonder that individuals who have withdrawn themselves from social media experience the same sort of withdrawal as from alcohol, nicotine, or sugar?[44] And that we now have social media rehab services? (See Chapter 14.)

It's Not What You Look at, It's What You See

Technology watchers, like MIT's Sherry Turkle, have argued that we have put ourselves on a digital diet: smitten with our technology, just like our processed food diet—and just as addictive. She argues that our digital diet has created a dissonance between empathy and compassion: there is a 40 percent loss of empathy in college students as a result of possessing a smartphone. In order to reclaim our contentment, we need to reclaim our capacity for solitude, which is undermined by our technology and our devices. Solitude isn't just being alone; it is a sense of self that is not derived from internet connectivity. The sentinel achievement of childhood is harnessing solitude and turning it into personal and spiritual growth. "If we don't teach our children how to be alone, then we doom them to always be lonely." No wonder comedian Louis C.K. won't let his kids have cell phones:[45] because the key social interaction, and subsequent social empathy, requires actual social participation. "You need to build an ability to just be yourself and not be doing something. That's what the phones are taking away . . . the ability to just sit there. That's being a person." Those are the children I see in my clinic today. They are the ones who prefer to text than talk, who can only communicate through Snapchat and won't make eye contact. They are the ones whose iPads have assured that they have never had, and never will have, a dull moment. And that's a whole lot of unhappiness.

17.

Contribute
(Self-Worth, Altruism,
Volunteerism, Philanthropy)

How many times have you heard the axiom "Money can't buy happiness"? Do you believe it? Do you think it can? Money certainly buys stuff, and stuff can certainly bring pleasure. But could money actually make you unhappy? Or do unhappy people just wish they had more money? Or both? If constant reward that begets even more reward-seeking behavior is the killer of contentment, then just maybe the more money you make, the more unhappy you get? Could money instead be buying us unhappiness, in the form of loss of contentment? And, if so, can it be reversed? Many philosophers have commented on the negative impact of our materialistic society on contentment in terms of home life, relationships, family, spirituality, and community. Princeton sociologist Robert Wuthnow notes that 89 percent of Americans agree that "our society is too materialistic." But when quizzed further, it appears that those 89 percent must be talking about the other 11 percent. In their own lives, they wish they had more money, a better home, and a faster car. Perhaps this is why we have seen so much political upheaval among those who identify with the middle class, who believe that our current economic system has left them behind. But this hasn't stopped them from spending, because they're convinced that without the latest gadget, they can't survive. The question is, does the misery of credit card debt outweigh the thrill of the newest iPhone? Madonna may have been the first Material Girl, but we are all living in her material world. Rather, I will argue that it isn't the money that ups your serotonin. It's what you do with the money and your time. Coco Chanel got it right: "There are people who have money, and there are people who are rich." Money can facilitate the contribution you provide to others.

Windfall Profits May Not Be as Profitable as You Think

Let's look at some people who have the opportunity to evaluate this axiom firsthand. Who goes from poor to rich quickly? Lottery winners. Do they become happy when they win the lottery? The lottery sure wants you to think so; otherwise, what's the point of buying all those tickets? The first evaluation of lottery winners was in 1978, where twenty-two lottery winners ($50,000 to $1 million) were compared with twenty-two controls (people whose financial situation had not changed), and also with twenty-nine accident victims who had become paralyzed (metaphorically losing life's lottery).[1] The lottery winners' happiness spiked when they won, while those of the accident victims' plummeted in the very short term. But over the next several months each group's level of subjective happiness returned to baseline levels. In fact, occasional studies describe financial ruin and even depression following a big monetary windfall.[2]

However, there are three problems with such studies. First, each examines different aspects of post-lottery life. Which definition of happiness are they using—and what if the examiners themselves don't understand the difference? As Kahneman and Deaton posited (see Chapter 12), are they measuring pleasure (that is, "life satisfaction") or are they measuring happiness (emotional well-being)? Second, how do you evaluate the winner's baseline state before they've won? Any study that requires looking backward must be taken with a grain of salt. Third and most important, who are these lottery winners, anyway? Are they onetime players who just got lucky, or are they chronic gamblers who were in it for the variable reward (see Chapter 14) and may keep jonesing for the next fix even after their metaphorical itch was scratched? Another way you could look at this is by observing who quits their job after winning the lottery. Turns out the majority of winners stay within their career paths, because that is where they derive their contentment through feelings of self-worth. They see their work contribution as meaningful.[3]

Check, Please

Maybe more money could buy you better food. Better food—high-quality protein with souped-up tryptophan and omega-3s and lots of fiber and even a sous-chef to prepare it for you. Maybe that would make people happier? In fact it could, but it doesn't. Analysis of eating patterns from the USDA Economic Research Service shows that as income increased from 1960 to 2013, the percent of money spent on food per capita decreased from 17 to 9.6 percent. Going further into the weeds, food consumed outside the home has increased from 26 percent in 1970 to 50.1 percent in 2014[4] . . . so we do have a sous-chef! But eating out doesn't necessarily mean eating well, even if you think you are. (Does the soda come with free refills?) Check out the dressing ingredients on your next Chinese chicken salad for proof. Furthermore, the lowest-income quintile spends $4,000 per year on food (36 percent), while the highest-income quintile spends $11,000 per year (8 percent). Altogether, the U.S. spends 6.7 percent of its GDP on food. Compare this to the French and the Japanese, who spend 14 percent of their GDP on food. But when you look closely at what we buy, it's not eggs, meat, or fish; rather, it's corn, wheat, soy, and sugar—everything that's subsidized.[5] In other words, America buys a whole lot of more pleasurable food for less money than the rest of the world, but in general, with the exceptions of some people on the two coasts, we're not buying better-quality food.[6]

Consumer Reports It's a Lemon

Psychologist Tim Kasser[7] of Knox College has spent his career trying to tease out the answer to whether a materialistic mind-set results in its own negative consequences. In developing a psychological tool called the Aspiration Index, Kasser identified four aspects of well-being in young people: self-actualization (being comfortable in your own skin), vitality (energy, alertness, feeling alive), depression, and anxiety. It didn't matter whether he assessed adolescents,[8] U.S. college students,[9] or Gen Xers from other countries;[10,11] the same pattern kept appearing. Those who valued financial success appeared to derive less contentment from life. This negative association between "material goods" and discontentment even held up under a rigorous meta-analysis[12] (the more you have, the more you want), while "aspirational life goals" correlated positively with personal contentment, even after controlling for financial status.[13] In other words, working for your own personal benefit is the reward for your labor. But it is the impact of your work or your actions beyond yourself—how you contribute to the greater good—that translates into contentment.

Could people grow out of this materialistic mind-set? Would it make a difference? Kasser has studied young people in the U.S. and Iceland over time, for six months to twelve years.[14] Guess what? If their affinity for materialism increased, their subjective well-being worsened, and vice versa. So the association holds up over time. Important, but that still doesn't prove cause and effect. Does materialism erode contentment, or, rather, could the change in contentment come first, driving materialism as a sort of haute couture booby prize?[15]

To determine causation, you need to design an intervention that either increases or reduces levels of materialistic thinking in order to see whether emotional well-being changes in the process. One set of investigators did a simple experiment: they divided some college students into two groups; the first group was told they were "consumers" and the second that they were "individuals." Surprisingly, the consumers responded to various presented consumer cues with materialism, selfishness, lack of coopera- tion, and lack of social contact, while the individuals exhibited none of these traits.[16] Apparently just being a consumer generates negative vibes.

The takeaway from all of these examples is that wealth can exaggerate your current situation, but it can't fix it. Depression itself is not dependent on income, although the ability to seek and pay for treatment may well be. If you're unhappy, with money and surrounded by people you don't trust, money will only make those problems worse. If you're fulfilled, have enough to pay for necessities, and enjoy a life of strong relationships, money won't actually make you more content, because the serotonin effect is not driven by wealth or income. Not in individuals, and not in countries. Ben Franklin said, "Money never made a man happy, nor will it. There is nothing in its nature to produce happiness. The more of it one has, the more one wants."

A Penny Saved Is a Penny Learned

Now it's time for you to do your own thought experiment. Don't consider yourself a consumer, powerless to the whims of the next fad and latest mar- keting scheme. Contrary to what Las Vegas says, you're more than your money. Think of yourself as an individual with unique morals and values, and who provides your own unique contribution to your family, to your work, and to the world in general. Think of yourself as an individual with superpowers unrelated to your wallet or your bank account. How does this make you feel? Any empowerment? Before buying the next product that you just have to have that is going to make you happy, engage your PFC first. Visualize yourself with this product three months into the future, six months, a year. Did you need it in the first place? Will it still make you happy? Will it help you to live your life in a better way? When you consider that product's worth in this way, do you still need it? My cousin has an entire shelf of anti-wrinkle creams. Each purchase, she swears, will make her look younger and therefore happier. Now, if that face cream could improve her self-image to the point where she would be less self-conscious and more social, that would be a product worth the expenditure. Yet most of them haven't even been tried before the next one is purchased.

Kasser also showed how you can dissociate consumerism from capitalism. He developed a computerized financial education package designed to get kids to save rather than spend money.[17] He randomized a bunch of adolescents to participate in a three-session intervention designed to decrease spending and increase sharing and saving, while another group served as the control. Both groups were followed up every six months for one year. Those who received the intervention demonstrated lower levels of materialism and higher scores on self-esteem (although anxiety remained unchanged). These various studies, while not proof, nonetheless argue that money is only worth the contribution it helps people to achieve.

I'll Show You Who's Boss

Contribution is the key concept here. Does contributing to the greater good or to society at large drive contentment? Can you derive contentment from meaningful work? Sociologists have wrestled with this for years, and the picture has now come into focus.[18] There are jobs that make you feel good about yourself and others that destroy your self-esteem. If your job or your boss (1) disconnects you from your values (for example, pitting the bottom line against the quality of work), (2) takes you for granted, (3) requires pointless or redundant work, (4) treats people unfairly, (5) overrides your better judgment, (6) isolates or marginalizes people, or (7) puts people in harm's way (physically or emotionally), then you have a job that will generate significant unhappiness. You come home stressed and exhausted and head straight for the chocolate cake or the liquor cabinet. Note that both sets of job characteristics that impact your mental health are exclusive of salary. But if you're one of the lucky ones, you experience your job as (1) self-transcendent (i.e., it matters more to others than it does to you), (2) poignant (challenging at difficult times), (3) episodic (with peak experiences that vary), (4) reflective (you can see the role that the completed work product will have on society), and (5) personal (you are proud to have performed it), then you have a job that can provide both life satisfaction *and* contentment. Note that both sets of job characteristics that impact your mental health are exclusive of salary. No wonder unhappiness at work is so rampant, in part due to job stress, and has worsened over the last three decades, from 40 percent in 1987 to 52 percent in 2013.[19] As the saying goes, "Find a job you love, and you'll never work a day in your life." Until your position, your mandate, your location, or your boss changes.

A friend of mine spent years as a guard in a maximum-security prison. She got a great paycheck but suffered immense stress and lack of sleep, and eventually became addicted to pills. Now she works in the floral department at a supermarket. She earns significantly less in dollars but has gained immense life satisfaction in creating arrangements and making customers smile. She sleeps better, has kicked her pill habit, and is much happier. Not all of us can quit jobs we hate, and not all of us would find inner contentment in cutting and wrapping flowers for little more than minimum wage. But what if you volunteered to do so or to create/ participate in something that brought joy to others?

On Moral Grounds . . .

Obviously, individual personal achievements such as doing well in school or getting a varsity sports letter are great ways to derive personal satisfaction, and will rack up points on Kahneman and Deaton's Life Satisfaction Index, but will those achievements drive contentment? It depends on what the underlying motive is. Altruism is the process of performing tasks that contribute to the "greater good" while deriving no personal gain or reward. Altruism doesn't activate dopamine but instead drives serotonin. Many seemingly selfless acts aren't necessarily completely altruistic, as there may be rewards in such actions, like becoming teacher's pet, earning Boy Scout badges, receiving public service awards, tracking for faster promotion, etc. Rather, the question is: Does your achievement contribute to a goal—that is, a goal that's bigger than you, that involves others? The answer uses what is commonly referred to as "moral decision making"—processes that govern thoughts of interdependence, egalitarianism, justice, charity, and empathy—versus those that govern thoughts of independence, aggression, punishment, and callousness. Is moral decision making anatomically and biochemically driven? There are two major classes of moral decision making, and our old friends the PFC (our Jiminy Cricket), serotonin, and dopamine take center stage yet again.

Altruism vs. Spite

First, have you ever cut off your nose to spite your face? Have you ever punished someone even though there were negative consequences for you as well? To wish others—and yourself—harm is the ultimate expression of lack of contentment. Known as *altruistic punishment,* or "spite" for short, this behavior stems from an impulsive and emotional reaction to what is deemed to be extreme unfairness. There are two routes to being spiteful.

1. The first route to spite is to have a dysfunctional PFC. Psychologists examine this area of the brain by playing the Ultimatum Game, which is a popular experiment in behavioral economics. One person is the "Decider," the other, the "Responder." The psychologist offers the players $100, conditional on whether the Decider and the Responder can agree on the split; otherwise both get nothing. The Decider proposes whatever type of split he or she thinks is equitable, and the Responder has one chance: Agree or get nothing. If the Decider offers the responder 10 percent, and the Responder thinks the offer is petty, the Responder may well reject it, even if that means both players get nothing. Responders with damage (e.g., head trauma) to their PFC can't see the benefit of anything but an even split; they have reduced capacity for compassion, shame, and guilt, and reduced frustration tolerance. So it shouldn't be surprising that in the Ultimatum Game they invariably get nothing. They can't see the prize because they can't see their spite.[20] Also, people with lesions to their PFC can't differentiate honesty from self-interest.[21] How happy can you be when you're willing to screw both yourself and others, you can't tell the difference, and you can't even help it?

2. The second route to spite is to lower brain serotonin levels. Normal volunteers played the Responder in the Ultimatum Game twice, either after consuming a tryptophan-depleting drink (which lowers serotonin levels in the brain; see Chapter 7) or a control beverage.[22] During the tryptophan depletion, Responders couldn't accept a deal;

they showed increased impulsivity and vindictiveness, which was predictable based on the change in tryptophan levels in their blood. This shows that reducing the molecule of contentment biochemically resulted in more impulsivity and spite behaviorally. Even giving tryptophan to an ornery dude can acutely improve his mood.[23]

A Benevolent Brotherhood of Man

The second type of moral decision making is called *aversion to harm*, either to yourself or others. Most people would rather inflict pain on themselves than on others, in what is known as "hyperaltrustic" behavior. In a set of experiments, investigators looked at whether taking even just one dose of the SSRI citalopram (to increase serotonin) or the dopamine precursor L-DOPA (to increase dopamine) could alter this behavior.[24] The subjects didn't notice that they felt any different, but their behavior was. The investigators rigged a contraption that would unleash a series of shocks either on the subject or on an unwitting victim. The subject was the Decider in that they could hit a button or not, but half the time (randomly) they ended up shocking themselves, while the other half the time they shocked the victim. Only the Decider was paid, and received different amounts to hit the button or not. More shocks meant more money, while fewer shocks meant less money. The investigators found that in the presence of the SSRI citalopram, the Deciders increased their harm aversion while strengthening their hyperaltruism: avoiding harm to others meant more to them than the money they earned. Conversely, with L-DOPA, hyperaltruism was reduced, which meant that the money must have been more important than any remorse at shocking others (i.e., they turned a nice guy into an a—hole). More recently, investigators specifically targeted dopamine in the PFC (our Jiminy) by administering a drug that knocked out the enzyme COMT (their Pac-Man in charge of dopamine clearance; see Chapter 3), and then had them play the Dictator Game: similar to the Ultimatum Game, but the Responder has no say at all. By increasing dopamine in the PFC with the drug, the desire to be equitable increased. Even though they had nothing to gain, they still wanted to level the playing field. So it goes both ways. Our moral decision-making capacity is biochemically determined and potentially subject to influence, depending on where and how dopamine and serotonin are acting. Then bludgeon your Jiminy—inhibit the function of the PFC through stress, sleep deprivation, or psychoactive drugs, even temporarily—and the full moon will awaken the werewolf in any of us. Bottom line: contentment and altruism co-migrate, as they are both dependent on serotonin; while reward and spite co-migrate, as they are both dependent on dopamine. Change the neurochemistry—change the emotion—change the behavior.

Property Management

How do these moral decisions play out in real life? You just can't go around giving people drugs and shocking them, but it's not all that hard to determine whether people are cooperative with each other, in order to achieve the greater good. One seminal study looked at the psychology of almost 1,200 Han Chinese farmers based on one variable: Did they live north or south of the Yangtze River?[25] North of the Yangtze they grow wheat, have their own plot of land, and are independent of other farmers and their travails; individuals within their societies demonstrate independence ("I'd rather die my way than live yours"). South of the Yangtze they grow rice, in paddies, where the water level is crucial. Since water runs downhill, the water level is not in your control. A drought or a flood in your paddy means a drought or a flood in everyone else's. So the rice farmers have to band together and work for the greater good; thus individuals have become interdependent so that everyone is uplifted to-gether ("a rising tide lifts all boats," literally). Interestingly, the rice farmers demonstrate signs of East Asian holistic thinking and of contentment—e.g., greater loyalty and lower divorce rates—whereas the wheat farmers show more signs of Western reward thinking—e.g., rugged individualism and higher divorce rates.

The Humanitarian Award

So can we change our brain's biochemistry? Of course we can, and with-out drugs. Jeez, that's what this whole book is about. Although we don't have the hard neuroscience behind it, one easy way to increase content-ment and derive health benefits is through volunteerism. By offering your spare time to a cause bigger than yourself, without thought of personal gain, you can derive meaningfulness and contentment and eudemonia. Several mechanisms may explain the association between volunteering and emotional well-being. Those who volunteer have a larger face-to-face social network (see Chapter 16) and more opportunity to derive a sense of contribution and purpose.[26] Physiological effects include reduced blood pressure and heart rate, suggesting reduction in anxiety or stress. A re-cent meta-analysis showed that volunteering improved depression, life satisfaction, and well-being, as well as resulted in a 22 percent reduction in risk for death.[27] And a recent analysis of a large UK population survey corroborated improved mental well-being in middle-aged and elderly populations that volunteer.[28]

Perhaps volunteering can even change adolescents for the better. In a randomized study of Canadian high school students, those who volun-teered to tutor elementary school kids for four months demonstrated lower BMI, lower inflammatory markers, and improved cardiovascular risk factors.[29] Analysis of the intervention group suggested that those who

increased the most in empathy and altruistic behaviors and who decreased negative mood exhibited the greatest decreases in cardiovascular
risk over time. So making the world a better place also tends to make a
better you.

The Benefits of Beneficence

If you don't have the time to make the world a better place yourself, pay
someone to do it for you. It's not the same thing, but it still works. Winston Churchill, brought up with all the advantages money had to offer,
famously said, "We make a living by what we get, but we make a life by
what we give." President George H. W. Bush in his convention's acceptance speech in 1988 spoke about the "thousand points of light," implor-
ing Americans to take up charitable causes. The good news is that among
young adults ages twenty to thirty-five, 75 percent gave something to
charity last year, indicating that while they don't volunteer their time,
they still do subscribe to the notion of "making the world a better place."
In fact, philanthropy is a way to use wealth for achievement, to transcend
oneself for the greater good. Then it should be no surprise that philanthropy is regulated by the same brain areas and neurotransmitters.

Harvard psychologist Mike Norton likes to give away money to people who will also give it away.[30] When research subjects are told to spend
their "experimental charity" on themselves, their happiness index barely
moves a notch. Yet when they are told to give that money to another person (prosocial spending), their happiness increases by the amount they
gave. Of course, the question is whether they would have felt the same
giving money out of their own bank accounts instead of Norton's.

Early brain scanning studies demonstrated that the reward pathway
(NA) as well as the PFC both light up in response to either taxation (taking your money involuntarily) or donation (giving your money will-

ingly),[31] suggesting dopamine might play a role in both. However, these pathways have now been further teased out. Activation of the PFC appears to be the arbiter between whether a potential donation is considered altruistic or charitable as opposed to offensive or cloying.[32] What role does serotonin play? In a small but ingenious experiment,[33] thirty-two European students were randomly assigned to receive either an oral dose of tryptophan or a placebo. After rating their mood on a visual scale and doing some unrelated diversionary tasks, they were given ten euros for their participation. At the exit there was an opportunity to donate to one of several charities. Upon departure, those who received the tryptophan donated twice as much as those who received the placebo. Obviously, this is a small study and does not prove causation. Does this work in the opposite direction? Can giving make you happier? We don't know for sure—but why don't you "give" it a try?

18.

Cope
(Sleep, Mindfulness, Exercise)

Life is going to throw you fastballs, curveballs, screwballs, and every so often you're going to get hit by the pitch. Despite your best efforts, your candidate loses, your kid gets sick, and you don't get hired for that ideal job. So how do you cope without being a curmudgeon or throwing in the towel? I know you've heard it all before: Eat right, exercise, get enough sleep, and breathe. First, what do any of these things mean? How much is the right amount? Second, it's not just about reducing your eye puffiness or fitting into those Lululemon leggings. There's hard science behind it.

One of the primary drivers of reward (Chapter 4) and inhibitors of contentment (Chapters 7 and 10) is stress. Yet it's not the specific stressor that matters, it's the individual's response to stress and how long it goes on that determines whether that particular stress is adaptive or maladaptive. For instance, most people would view studying for a test or running the one-hundred-meter dash in the Olympics as "good stress," in part because the stress is acute, there are positive benefits to be had, and you can see yourself being happy past the end of the event. Conversely, most people would view high demands at work or caring for a parent with dementia as "bad stress," in part because the stress is chronic, there are few if any benefits to be had, and there's no expectation of being happy— no light at the end of the tunnel. But chronic stress doesn't just wreak havoc on the individual; it costs everyone. A recent report from Harvard and Stanford Business Schools put the annual price tag of U.S. work stress at 120,000 lives and $190 billion.[1] Jazz vocalist Bobby McFerrin beseeched us: "Don't worry, be happy." Great idea, but what drives worry and how do you reverse it?

Throwing Jiminy a Life Preserver

Tamping down anxiety is the function of the prefrontal cortex (PFC), your own internal Jiminy Cricket; yet it also tamps down the experience of reward. When it's working properly, the PFC will reduce amygdala (the fear center) output (*I don't need to fear this*) and increase hippocampus (the memory center) function (*I've been here before*), focus you, keep you to the task at hand, and reduce hypothalamic activation to maintain low cortisol levels, which keeps your metabolic function stable. But when the PFC is worn out or damaged from years of chronic stress, as can happen in communities where the population is food insecure (not knowing where the next meal is coming from, e.g., in Memphis or Sudan) or life insecure (fear of being in the cross-fire, e.g., in Chicago or Istanbul), then cognitive control is disinhibited and impulsivity is let loose.[2] A dysfunctional PFC means less restraint on the reward pathway, with the prospect for non-stop reward seeking for ever-elusive pleasure and ultimately an increased risk for addiction. That's not all: a crippled Jiminy also means an increase in cortisol, a suppression of serotonin-1a receptors, and an increased risk for depression (see Fig. 10-1). Nurturing our PFC should be our prime directive; unfortunately, our environment has claimed our PFC as collateral damage. We have three simple methods to give our PFC the rest it needs—sleep, mindfulness, and exercise—but unfortunately none of these are simple in modern society, although each is crucial to your physical and mental well-being. Let's deal with them in turn.

Sleeping Your Way to the Top

As we showed in Chapters 9 and 10, sleep is essential to optimizing serotonin and mood. The brain after a good night's sleep processes information differently than a sleep-deprived brain, with decreased activity in the amygdala and increased connectivity to the PFC.[3] Consistently getting a good night's sleep corresponds with beneficial changes in the brain, including your memory centers and your Jiminy.[4] Conversely, chronic sleep deprivation results in increased dopamine and reduced serotonin, while the increased psychological stress and cortisol of sleep deprivation will decrease serotoin-1a receptors. Stress impedes duration and quality of sleep, which leads to more stress. And round and round we go.

Sleep deprivation takes its toll on your brain's capacity to function. In one five-day study,[5] adults were randomly selected to be in one of four groups: normal workload + 8 hours of sleep; normal workload + 5 hours of sleep; excessive workload + 8 hours of sleep; and finally excessive workload + 5 hours of sleep. As you might expect, increased workload led to fatigue and sleepiness regardless of sleep duration, but did not alter quantitative work performance or wakefulness tests. Sleep restriction on its own led to a worsening of all cognitive tests. Perhaps not surprisingly, with increased work demands *and* less sleep, subjects performed markedly worse on all tests, and with actual changes in brain activity along with adverse brain metabolic changes, especially in the PFC. Sleep deprivation also takes its toll on the immune system. If you give a group of a healthy adults a cold virus and isolate them for five days, which factors influence whether or not they get sick? Sleep. Those who slept less were five times more likely to develop colds.[6] Could increasing sleep duration turn this around?

At the present, 35 percent of Americans get less than seven hours of sleep per night (optimally you should get eight), and clinical insomnia (they can't fall or stay asleep) befalls 23 percent of the adult population.[7] Using eight hours as our benchmark, adults maintain a sleep debt of sixty to ninety minutes per night, which is not recompensed by sleeping in on the weekends.[8] In fact, your brain has an internal clock, and it doesn't like to be unwound; it prefers that you go to bed and wake up at the same time every day. Work environments generally value those who come in early and stay late, and those who respond to e-mails at 11:00 p.m. And yet, this can actually worsen an employee's job performance and overall well-being. In fact, sleep deprivation causes an average of 7.8 days per year in lost work performance, and costs employers about $2,280 per capita, or a total of $83 billion per year in the United States.[9] Outside the U.S., employers and governments have picked up on this statistic; most recently France has passed a law that prohibits employees from checking their work e-mails after 5 p.m. and on weekends.[10] In contrast, many Americans are perpetually glued to their e-mail during dinner, in traffic, and even while picking up their kids from day care. One enterprising Texas day care put a sign on their door: "GET OFF YOUR PHONE!!! YOUR CHILDREN ARE GLAD TO SEE YOU! ARE YOU HAPPY TO SEE YOUR CHILD?"[11]

Most people, when faced with the adverse consequences of their sleep debt, would opt for a quick nap. But unfortunately, while naps acutely improve cognitive functioning, they do not repair the negative mood of chronic sleep deprivation.[12] And most people don't take naps: they drink coffee or take 5-Hour Energy drinks. The caffeine will worsen their chances for restful sleep, and they get the added bonus of kicking their dopamine reward system into overdrive. Let's see you climb down now.

Not sleeping will seriously "harsh your mellow," thereby skewing your serotonin and any possibility of your achieving Zen. A good night's sleep is not a luxury; it's a necessity.

The Sounds of Sleep

One of the most significant causes of sleep debt is known as obstructive sleep apnea (OSA). If you're one of the 35 percent of Americans with sleep debt, you might well be suffering from it. In OSA, the airway collapses while asleep, preventing oxygenation and carbon dioxide exchange ("I can't breathe . . ."). At its extreme, OSA can irreversibly damage the right side of the heart. It can even kill you, albeit slowly and over years. People normally associate OSA with obesity, and for good reason. The fat around the neck contributes to the collapsing of the airway; and OSA stimulates the hunger hormone ghrelin, causing increased appetite. So OSA and obesity form a vicious cycle of weight gain and metabolic dysfunction, with resultant overeating and depressed mood. But that's just scratching the surface. Lots of normal-weight people also get OSA, possibly due to reduced muscle tone in the neck closing off the airway, and are at just as much risk for metabolic problems.

The best way to figure out if you have OSA is to ask your bed partner if you snore. I promise, they won't lie. The next best way is to record yourself when you sleep. And one final way is to notice if you wake up with a headache not related to how many cocktails you consumed the night before. If you have OSA, what are your options, and will they work?[13] The standard mode of medical therapy is continuous positive airway pressure (CPAP): a machine that blows air into the airway with a form-fitting facemask. In one study of three hundred women who suffered from sleep apnea, treatment with a CPAP machine not only helped them to sleep but they also reported a better quality of life, improved moods, better PFC function,[14] and decreased levels of anxiety and depression.[15] In insomniacs (people who have a hard time falling asleep and staying asleep), cognitive behavioral therapy can also improve clinical depression.[16] But CPAP is uncomfortable and noisy, which actually makes getting to sleep harder until people get used to it. Nonetheless, these studies show that when your sleep improves, your mood also improves, and your serotonin is moving in the right direction.

What about the rest of us who can't seem to shake the impending doom of tomorrow? Practicing good sleep hygiene can help. Unplugging all electronics in your bedroom is one of the best things you can do for sleep quality. Tossing and turning? The remote is your enemy! No binge-watching Netflix in your bedroom. No TV in there at all. Blackout curtains can help. Understandably, you may not be able to unplug *all* the electronics in your bedroom (e.g., your alarm clock). There are lots of things you can try to improve sleep hygiene: keeping your bedroom cool and dark, with plenty of fresh air; taking a warm bath or shower before bed; not eating or drinking after dinner; making sure to urinate before getting into bed; and, most importantly, no screen time for one hour before bedtime. This last one might be the most difficult for us to get used to. Many Americans deal with insomnia or try to wind down by bingeing on Netflix after checking their work e-mail, Snapchat, or Facebook. But the blue light emanating from the screen keeps your melatonin (the hormone in your brain that tells you when it's dark outside) from rising, causing a phase shift in your sleep cycle[17] and guaranteeing that you'll feel awful in the morning. Try picking up a book instead.

The Myth of Multitasking

What would happen if we did get enough sleep? Would our stress be lifted? Would our PFCs be any happier—and would we? Apart from sleep, the above studies suggest that there is an additive effect of workload and psychological stress on subjective well-being. Other studies certainly bear this out, as they point to reductions in PFC function and increase in cortisol levels.[18] No doubt these effects on the PFC make for diminished executive function (decision making) and increased amygdala activation of more cortisol, driving long-term metabolic dysfunction (e.g., diabetes) and immune system defects (e.g., inflammation).[19]

A lot of stress comes from our screens—and not just the blue light that gives us insomnia. Full-blown internet addiction is particularly worrisome and even life-threatening (see Chapter 14), yet many people suffer from a less severe but related problem, commonly known as "multitasking" (simultaneously processing multiple incoming streams of information), e.g., reading live tweets while IM'ing your sister as you watch the latest episode of *Game of Thrones*. While multitasking existed before the advent of the internet and mobile devices, the combination of these two technologies has shot this practice through the roof. Overall media use among America's youth has increased by 20 percent over the past decade, yet the amount of time spent multitasking with media increased by over 119 percent over the same time period.[20] Adults who engage in multitasking find themselves increasingly stressed. Many think that multitaskers are gifted people who are to be envied, as they are able to perform several tasks at once without compromising quality of effort or work performance. They are the pride of most companies, as they seem to juggle all the balls, keep the paper moving, take no vacations, and keep coming back for more—the overachievers that make you look like a slacker, and force you to measure up. I guess that's true, for the entire 2.5 percent of the population that can do this well.[21] Those imbued with this innate talent keep their PFCs whirring along despite dissonant information being pelted at them.[22] Yet multitasking, like sleep deprivation (which often go hand in hand), eventually takes its toll on the rest of us. In one cross-sectional study,[23] heavy media multitaskers were more susceptible to interference from irrelevant stimuli, were less able to screen out irrelevant information, and performed worse on tests of task-switching ability. And multitasking alters blood flow in the PFC as well as other information-processing areas associated with executive function.[24] Yup. If you spend too much time creating and retorting with memes, your Jiminy may get hindered. These results argue that multitasking is associated with deficits in fundamental information processing[25]—in others words, overachieving is accomplished with the help of smoke and mirrors, which may turn out to be a major contributor to our depressed mood all by itself. Multitasking, by increasing psychological stress, is associated with clinical depression[26] and might actually increase degenerative changes in the brain.[27] But, again, this is correlation, not causation. If you're a mere mortal, does multitasking cause changes in PFC function, and if so, are they reversible? Or are people with PFC dysfunction more likely not to sleep and multitask instead—say, starting a tweetstorm at 3:20 a.m.? We're still not sure.

Nonetheless, cutting down on multitasking to improve brain function and mood has been catching on. The easiest strategy is: TURN OFF YOUR PHONE! ALL CAPS, IDIOT! (Why do you think France banned work e-mails after hours?) Try it. Try to live without your cell phone for five days. The angst that knowing the phone is there but you can't turn it on or e-mail—the tremor of your hand, the sweat on your brow, the feeling as though a limb has been removed—is more than enough to demonstrate phenomena of both tolerance and dependence, consistent with addiction. Professional blogger Andrew Sullivan beautifully described his descent into cyber-Hades in his article "I Used to Be a Human Being" in *New York* magazine.[28] In the process of rehabilitation, he rediscovered his sense of self. Of course, the problem with doing this is that some of us actually do need to be connected in order to perform our jobs, pay the bills, and keep track of our family members. Worse yet, voice command recognition technology (e.g., Siri, Alexa) has eliminated vast numbers of service jobs around the world, and will continue to do so, which means you can't disconnect even when you want to. Nonetheless, taking the time to unplug, even for only an hour a day, is one of the most rewarding things you can do, as it's the path to reduce both dopamine and stress, improve sleep, and hopefully allow your serotonin to do its job.

Don't Just Do Something, Stand There

Another bona fide method to engage your PFC in a meaningful way and give your emotional brain, or the amygdala, a rest is meditation, which is becoming more and more popular everywhere, from the classroom to the boardroom. Meditation has been around for millennia and is practiced in many cultures and religions. For instance, Buddhist meditation practice, where you rid your mind of extraneous thoughts—a ritual closely aligned with attainment of spiritual happiness—has been shown to prospectively increase PFC activity, healing your Jiminy.[29] The most recent spin is a non-denominational practice called mindfulness, and we know more about its effects because we've been able to actually examine the brain before and after mindfulness practice in an MRI scanner. Based on Buddhist tradition, this technique was adapted by Dr. Jon Kabat-Zinn to help patients deal with chronic pain, and has since become a mainstream method for reducing the effects of stress on the body and the brain.[30] The basic premise is to "live in the moment" by learning to observe your thoughts and emotions dealing with the past and future without acting on them. By focusing strictly on the present, recognizing that no one is in control of either the past or the future, you act like a scientist and observe the natural phenomena within your own mind—leaving you in a state of greater peacefulness. A favorite exercise is to eat one raisin—very, very slowly—over about ten minutes. You roll it around on your tongue, explore the rough surface, bite into it in slow motion—and all of a sudden, it isn't a raisin. By doing so, stress and distractibility diminish. It's a new experience that you've never had before—at least, that's the concept, ostensibly an enjoyable one, although not everyone agrees (see the Meditation for Real Life series in the *New York Times*).[31]

Mindfulness sounds very New Age (kind of like a brain colonic), but there's a lot of neuroscience that goes with it,[32] including research showing that meditators have clear differences in brain structure. A meta-analysis of studies comparing meditators and non-meditators demonstrates changes in the brain related to the size of the frontopolar cortex and insula (which governs body awareness), the hippocampus (memory), the corpus callosum (transmission between both halves of the brain), and perhaps most importantly the anterior cingulate cortex and prefrontal cortex (the Jiminy Cricket, those areas involved in executive function that keep you from doing stupid things).[33] Of course, we are still left with the cause-and-effect question. Does meditation help your PFC? Or do people with better PFCs choose to meditate? Or both? We just don't have long-term longitudinal data as of yet. However, one study in veterans with post-traumatic stress disorder, as well as traumatic brain injury, did find beneficial effects of mindfulness-based training that lasted three months post-training,[34] and mindfulness-based training increases neural connectivity in regions known to play a role in empathy.[35] So you feel better and are generally a nicer person. Win-win.

Like any skill, it takes daily practice to master mindfulness. Those who practice it swear by it, but it's definitely not something you just dabble in. I am a pretty anxious guy myself, so eight years ago I took the mindfulness-based stress reduction (MBSR) course at the UCSF Osher Center for Integrative Medicine here in San Francisco. I personally found the experience quite valuable, even though I don't practice it every day like I should. I learned that anxiety is "excitement about the future." The problem is that the future never comes, and it is almost always worse in my head than it is in reality. Today I would worry about what might happen tomorrow; then tomorrow would come, and I would be focused on the day after that, and so on. So you're never in control. This is a great way to make yourself miserable. By focusing and spinning out on what might or might not happen tomorrow and what I had done yesterday, I was missing out on the fun and enjoyment of what was happening today. By always focusing on the future and the past, I generated my own chronic stress, and I was always missing out on the happiness of today.

This realization was a revelation, a lightbulb going off in my brain, and it's part of the reason why I wrote this book. Whenever my thoughts take me to tomorrow or yesterday and what the future may or may not hold (always leading to anxiety), I instead refocus my attention on what is good (and of course bad) that is happening today. And my self-assessment is that today, right now, is usually pretty darn good. I find that I am much calmer, I fidget less and I am much less distractible. I also stopped eating when I am not hungry, as I am not eating for stress anymore. I revel in today. And tomorrow? Tomorrow will be here soon enough and will take care of itself. I downloaded a stress-reduction app on my iPhone. I have found the best time to use it is when I am on a plane, after pulling away from the gate, and waiting in line for takeoff. I used to feel angst about the wait . . . not any longer.

Try it . . . right now. Close your eyes; feel the sensations around you. What do you hear? What do you sense? Do you feel your feet pressing into the floor? Your hands on this book or computer? What do you smell? Focus on right now. Not "What do I need to do tomorrow? What will I have for dinner? What will we do this weekend?" Try this for five minutes and focus on a saying. . . . It can be any saying: "I am here." "I have enough." "I am content." "I am safe." For these five minutes everything other than your breath is a distraction. Can't think of a good mantra? Be thankful . . . for five minutes. Make a "gratitude" list of things and people you are thankful for. Harder than you thought, isn't it? Try it for five minutes a day for thirty days, and see if and how it changes you.

Mind over Matter

My UCSF reward-eating team has studied the effect of a twelve-month mindfulness intervention added on to a standard diet-exercise program for obese patients along with a six-month follow-up period.[36] Half received just the diet and exercise intervention, while the other half received mindfulness training and the instructions to practice every day for about forty minutes. We were mildly surprised that the mindfulness subjects lost only a small (and not significant) amount of additional weight compared to the control subjects. But the more they practiced mindfulness, the fewer sweets they ate, and the better their blood glucose control, suggestive of improved metabolic health.[37] However, when we looked at specific fat depots, we were encouraged to find that the mindfulness group lost a lot of visceral (big belly) fat, while their subcutaneous (big butt) fat remained essentially the same.[38]

Visceral fat is not like any other fat in your body. While subcutaneous fat accounts for anywhere from 5 to 45 percent of your body weight, visceral fat only encompasses 4 to 6 percent. So when you stand on a scale, which are you measuring? You don't know. Subcutaneous fat, for the most part, is not metabolically active, and once formed is extremely difficult to eradicate. But excesses of subcutaneous fat do not contribute to poor metabolic health, other than the fact that people who don't like their bodies experience psychological stress because of it. In fact, subcutaneous or "big butt" fat can actually be protective in some cases. Conversely, it's the visceral fat that is the driver of the diseases of metabolic syndrome and depression; this relationship has been well documented in adolescents[39] and in adults.[40]

In our UCSF adult mindfulness study, virtually all subjects improved their metabolic health significantly. Fasting insulin, glucose, triglycerides—all reduced over the twelve months and stayed reduced even after the intervention was over.[41] These data suggested that mindfulness meditation reduced visceral fat, which in turn improved various health parameters. While no prospective study has yet been done, it's likely that mindfulness would prevent these metabolic problems from occurring in the first place.

How do we know your propensity to accumulate visceral fat is not just genetic? Well, to factor out other influences, you would have to look at identical twins. A recent study from Finland took ten sets of identical twin adult males who had the same weight and BMI as their twin, the same eating patterns, the same living conditions, etc. The only thing they differed on was leisure-time physical activity. One twin was active, the other a couch potato. The researchers evaluated all their metabolic parameters, all their calories consumed, and finally all of their fat depots.[42] In each case, the inactive twin carried about four extra pounds of visceral fat, probably the reason that the inactive twin weighed four pounds more than the active twin. And it was this visceral fat that correlated with their cardiovascular fitness and their fasting glucose and insulin levels. This study shows us clearly that inactivity is associated with increased visceral fat—exclusive of energy intake or genetics or family background or upbringing. And it's the visceral fat that predicts future metabolic disease. Visceral fat is the most malleable fat depot in the body: it's the easiest to lose. And it's the visceral fat that is directly amenable to exercise.[43]

Exercise Sculpts Your Brain as Well as Your Body

Doctors have known for decades that exercise is the single best thing you can do for yourself, both physically and mentally. Everyone thinks exercise makes you lose weight, yet there is not one study anywhere in the world's literature that shows that exercise alone causes weight loss: it causes visceral fat loss, but it also causes muscle gain, so the two tend to cancel out, and sometimes body weight even goes up.

The question is: Can exercise treat major depressive disorder? Can it make non-depressed people happier? Many people have heard of the endorphin rise with significant exercise, or runner's high. Does it open the gateway to happiness as well? Many prospective trials have now been conducted, and the overwhelming majority demonstrate that exercising is better than not exercising; exercise is about as good as SSRIs are in treating depression; and exercise + SSRIs is better than SSRIs alone.[44] And it doesn't matter what kind of exercise; cardio and resistance training both work. One explanation for these findings is that we know that stress increases cortisol, and that cortisol corresponds with decreased cell birth in the hippocampus. We aren't exactly sure how or why, but greater numbers of cells in the hippocampus is associated with happiness. Fewer cells equals more chance for depression. Exercise both increases the birth of these cells and offsets stress-induced cell death. One of the reasons we think antidepressants work is that they also correspond with new cells being born in the hippocampus. If we somehow negate or squash the growth of these cells, antidepressants don't work at all.[45]

Perhaps the most fabled benefit of exercise is the aforementioned runner's high, the sudden onset of euphoria experienced by hard-core marathoners. This phenomenon has been attributed to the release of the EOP beta-endorphin,[46] and more recently we've learned that the added benefit of sedation and alleviation of anxiety is due to the simultaneous release of endocannabinioids—our own personal hash stash present throughout the brain—[47]yet without giving us the munchies.

Well, that's great if you're a marathoner. But does exercise work in mere mortals who may be unhappy but not necessarily clinically depressed? Here the data are harder to come by. In large analyses with many subjects, there appears to be a significant effect of exercise on improved mood.[48, 49] Even adolescents, who are by nature moody, can benefit from exercise's effects on mood and depression.[50] (Of course, with some kids being perpetually glued to their smartphones or World of Warcraft, they're getting less exercise now than in years past.) And in a meta-analysis of 1,500 elderly people, exercise was found to be effective in decreasing symptoms of depression.[51]

But these are research studies . . . and people don't live in a vacuum. Weather, temperature, wind, and elevation all impact on one's desire and performance of exercise. A group in China took this on by geo-coding (using these geographic variables) to compare levels of happiness in twenty-eight countries around the world.[52] They also adjusted for GDP (see Chapter 12). Their findings showed that, after having adjusted for all the confounders, physical activity correlated with well-being, and lack of physical activity with the greatest unhappiness.

The effects of both meditation and exercise are real, but likely not enough to turn depression into joy. What about combining the two? Might there be additive effects? One short-term study paired both together; first subjects engaged in a forty-minute mindfulness practice, and then they got on a treadmill for another forty minutes, for eight weeks. The results showed that the combination of the two was better at alleviating depression than was either one alone.[53]

We shouldn't be surprised that virtually any stimulus that increases psychological stress also inhibits PFC functioning and can ultimately foster addiction, and any stimulus that specifically increases visceral fat will increase the risk for depression. Conversely, anything that can attenuate either of these two phenomena can turn these negative emotions around. All of the above remedies are tried, tested, and guaranteed, or your money back (but since none of them costs any money, don't expect a refund).

There's an App for That . . . or Is There?

But here's where the technology industry will try to get its hooks into you. Many companies and digital app manufacturers have been proffering "personalized" health monitoring and programs in the name of "wellness." What is wellness, anyway? Most insurance companies define wellness as "the absence of illness," because for them it's about not paying benefits: if they don't hear from you, you must be well. Exercise trainers equate physical fitness with wellness. But what if you're physically fit but financially destitute? Or what if you're sleep deprived? Meditation coaches define wellness as a state of spiritual calm or lack of stress. But what if you're calm because you smoke a few joints or take a few drinks to make the world go away? Each person has his or her own definition of wellness. But wellness really means so much more, and contentment is at the top of the list.

These companies will sell you a wearable computer that will do everything from monitoring your step count to monitoring your blood pressure[54] to monitoring your blood glucose. They'll sell you a set of ready meals brought to your door, and buzz you when it's time to exercise and sleep, and determine for how long you engaged in both. And some people are changing their behavior because of them. There are more than forty thousand smartphone apps on the market devoted to health and fitness. They have digital apps that can monitor your emotional well-being[55] and that use techniques such as self-monitoring, providing feedback on performance, and goal setting. Some of these apps are "gamified" with badges and monetary rewards to increase compliance.[56] They'll monitor your mindfulness and provide reminders to walk. You can compete with your friends to get steps in. We're like Pavlovian dogs, trained to respond to the dings on our cell phones. Clearly a burgeoning cottage industry.

All of this sounds too good to be true. Because it is. Oh, yes, these companies and apps can monitor your every heartbeat and generate lots of data. But do they alter your health or well-being? A systematic review of twenty-seven randomized controlled trials of smartphone apps yielded modest evidence of efficacy, with only half showing benefits.[57] Those that improved health tended to do so in conjunction with other modalities (e.g., a trainer or coach). These studies were not very long, ranging in duration from one to twenty-four weeks, with a mean of ten weeks. Only half the studies monitored continued engagement with the smartphone app, and we know that app usage usually falls off at the four- to six-week mark,[58] because these apps have yet to learn how to turn data into information that the individual can use. And the lack of actionable data leads most users to eventually curtail their use after time.[59] These algorithms are just not ready for prime time.

Back to the Prime Suspect

Stress and sedentary behavior have been around for a while. Yet today addiction and depression are overwhelming public health problems. Now, at long last, we must deal with the most pernicious denizen of our Western environment and culture, the most toxic stress of all. The factor that causes more cases of addiction, depression, disease, and unhappiness than all of the others combined. The toxic brew to which all of America, and indeed, the entire world—old and young, rich and poor, Caucasian and African-American and Latino and Asian, educated and not—is now exposed. The toxic constituent that masquerades as our pal, our "homey," our BFF. The toxic item that has invaded our homes, our schools, our workplaces, and our bodies, and we willingly open the door: toxic food.

19.

Cook
(for Yourself, Your Friends,
Your Family)

Which brings us to our final question: How did all of these pathways change in the last forty years? Why are addiction and depression today the number five and six diagnoses, just behind the various diseases of metabolic syndrome (heart disease, hypertension, type 2 diabetes, cancer) that are numbers one through four? Maybe it's because people who think they aren't exposed to something actually are? And maybe that exposure is the same across the country and regardless of class? Or around the world? There are many causes, but the one we haven't yet addressed is the one that is currently affecting almost everyone in the world regardless of class. And what if that exposure is mixed into all of our food without our knowledge? And what if that exposure just happens to be addictive? What's the cheapest pleasure?

Of course, the answer is sugar—the other white powder. Now you're thinking, *Great, I made it through this entire book, and Lustig is bringing up sugar*—again! But, it's true. Sugar isn't just responsible for many of our physical health problems, but it also plays a significant role in our mental health. Let me prove it to you. Sugar is the stealth ingredient that's been added to virtually every industrial recipe to make processed food palatable, and ultimately saleable. As I described in Chapter 6, sugar fulfills the criteria of a substance of abuse, as it is toxic and addictive. It also meets criteria for regulation, as it is ubiquitous (can't get away from it) and detrimental to society.[1] If most of the food in the grocery store is spiked with added sugar, and you don't know, because there are fifty-six names for sugar on the label and you don't know them, then how do you avoid it? And if the fiber is removed from fruit to turn it into juice yet no sugar is added, can you even trust the label? And if its biochemical properties fry your liver, and its hedonic properties fry your brain and make you want more, how do you keep from succumbing?

Sweetening the Pot

The American Heart Association has argued for a daily limit of 6 teaspoons of added sugar for adult women, 9 teaspoons for adult men, 3–4 teaspoons for children,[2] and none for toddlers below the age of two.[3] The World Health Organization and the USDA are more lenient, declaring 12 teaspoons of added sugar per day the maximum. However, American adults consume an average of 19.5 teaspoons of added sugar per day, and children consume an average of 22 teaspoons per day. Latinos, African-Americans, and Native Americans consume between 25 and 50 percent more than their Caucasian counterparts.[4] These minorities are at greater risk for developing metabolic syndrome due to their added sugar consumption, especially from soft drinks.[5,6] And we know that these same minorities are higher risk for mortality when they manifest severe mental illness.[7] But of course this is all correlation, not causation.

Here's the conundrum: What if we decided to cut down our personal sugar consumption? What if we consciously removed sodas, candy, cakes, and ice cream—everything we normally call dessert—from our homes and from our diets? Turns out we'd still be over our limit, because only 51 percent of the sugar in our diet is in the foods that you'd expect. That means that 49 percent of the sugar we consume is in foods and drinks that we didn't know had sugar. Salad dressing, barbecue sauce, hamburger buns, hamburger *meat*, as well as so-called healthy options, like granola and muesli. And don't get me started on the health benefits of fruit juice, which is basically just sugar without the fiber. You're still at risk for diabetes.[8] So even if you cut out dessert, you'll still be over your limit, because of the rest of the sugar that's in processed food.

The industry argues vociferously that sugar is a required and necessary ingredient in their recipes. Here are a few of the industry's pro-sugar arguments, and why it's good for them and bad for you.

(1) Sugar adds bulk. Did you ever wonder why Lucky Charms has marshmallow stars, hearts, moons, and clovers? Because kids like them? Well, yes, but really because marshmallows are cheaper than oats. By taking up space in the box, the industry saves money on oats and can sell the box for more. A great business strategy.

(2) Sugar makes food brown. This is why bananas brown and why we paint barbecue sauce on our ribs on the grill. It's called the Maillard, or "browning," reaction. Well, that reaction is happening inside your cells all the time, and when it does, two things happen: proteins unravel, and free radicals form, which further damages cells. The Maillard reaction has another name: the aging reaction. Every time this reaction occurs, it throws off an oxygen radical, which is similar to hydrogen peroxide: it's great for killing bacteria on your skin, but it also kills liver cells, which is why so many people who have fatty livers progress on to cirrhosis. And fructose causes that aging reaction to occur seven times faster than glucose. Your body, and especially your liver, is aging faster with sugar. Just like it does with alcohol. Not good for your physical or mental health.

(3) Sugar raises the boiling point. This allows for caramelization to occur, which is very tasty, but again this is just the Maillard reaction, which, over time, can cause your cells to age. Now there are data to suggest that fructose could "caramelize" your hippocampus,[9] which could remove the brakes from your dopamine transmission, squashing your PFC, your Jiminy Cricket.[10]

(4) Sugar is a humectant (it attracts and maintains moisture). How soon does fresh bakery bread become stale? Maybe two days? How about grocery store commercial bread? More like three weeks. Ever wonder why? In commercial bread, the baker adds sugar to take the place of water, known as *water activity*, because sugar doesn't evaporate: it takes up space in the bread, and the sugar molecules hold on to water during baking, so the bread stays moist. Furthermore, the sugar will attract water from the air, so the bread won't dry out after it's baked. Good for the industry, bad for your health.

(5) Sugar is a preservative. Have you ever left a soda at room temperature open to the air? Of course, after the carbonation escapes, it goes flat. But do bacteria or yeast ever grow in it? Never.

The question is whether the hit to your liver provided by each dose of sugar is worth the eventual decline in physical health, risk for disability, and increased medical costs. The answer is in: sugared beverages alone account for 180,000 deaths per year worldwide, and for about 10 percent of all disability-adjusted life years.[11] And this is especially true in the twenty- to forty-five-year age group, which is experiencing rates of disability like never before.[12] And some of that disability is mental health, linked to sugar consumption.[13]

A Little Less Sweet

My UCSF metabolic team recently completed a study where we took forty-three children (Latino and African-American), aged nine to nineteen with metabolic syndrome, and who consumed at least 50 grams of sugar per day. We studied them on their baseline diet for various aspects of metabolic health and fat in different organs. We then catered their meals for the next nine days to have the same caloric content, the same percentage of protein, fat, salt, and carbohydrate as their usual diet. The only difference was that we substituted starch (glucose) for the added sugar (glucose + fructose) in their diet. We took the sugar in their diet from 28 to 10 percent of calories and kept everything else the same. We substituted bagels for doughnuts, baked potato chips for sweetened yogurt, and turkey hot dogs for chicken teriyaki. We didn't give them good food; we gave them processed food, but without all the added sugar. We gave them a scale to take home, and each day they weighed themselves. If they were losing weight, we counseled them to eat more, in order to keep them weight-stable. At the end of the nine days of eating our food, we studied them again.[14]

Guess what? Every aspect of their metabolic health improved significantly after just ten days of eating the lower-sugar meals. Their subcutaneous (big butt) fat did not change (after all, they had not lost weight), but their visceral (big belly) fat dropped and, more importantly, their liver fat dropped a lot.[15] Their blood lipids (markers of heart disease risk) all improved, and in just ten days.[16] We improved their metabolic health and quality of life without changing their weight. And most importantly, *they all felt better!* They had more energy, they could concentrate better, and anecdotally the parents said that they were less disruptive in class.

So what about mental health? Is the acute buzz from each soft drink worth the eventual negative impact on mood? In preschoolers, sugared beverage consumption correlates with behavioral problems.[17] In teenagers, sugared beverage consumption correlates with violence, severe depression, and suicidal thoughts.[18] Studies in Australia and China show that sugared beverage consumption correlates with unhappiness independent of chronic disease development.[19, 20] Furthermore, a study of pregnant Norwegian women found a direct relationship between sugared beverage consumption and loneliness.[21] Of course, these studies are all correlational, not causational. These data do not clarify whether soda can make you lonely, depressed, and violent, or whether lonely, depressed, and violent people are more likely to drink soda or eat Ben & Jerry's as their reward.

Stepping Down from the Food High

There are at least three reasons to eat: hunger, reward, and stress.[22] Within the SHINE study (see Chapter 18), our UCSF reward-eating team examined how obese women's reward system drives their food intake toward problem foods and whether stress heightens the reward response to specific foods.

We found that one-third of the obese subjects in our study reported times when they experienced a loss of control over their eating, and when they did, they gravitated to highly palatable (high-sugar, high-fat) foods. The question was: Could a mindfulness-based intervention reduce both their stress and their food cravings in order to improve their metabolic health and their mental health?

First, we needed to determine who these subjects were. Our UCSF reward-eating team developed and tested the Reward-Based Eating Drive (RED) scale.[23] We also found that reductions in reward-driven eating predicted the success of a mindfulness-based weight loss program.[24] Other studies have shown that obese people exhibit a loss of dopamine receptors (tolerance) in the reward pathway.[25] These data implicate both the dopamine and the EOP systems (see Fig. 2-1) in this type of reward-based eating drive.[26,27,28]

The next thing we needed was a biomarker or lab test of reward-based eating, something that would inform both the subject and us that the reward system was in overdrive and at the root of their eating behaviors. We already knew that some people who were exposed to a drug called naltrexone (which specifically blocked their EOP receptors; in order to block the consummation of the rewarding behavior, see Chapter 3) would often become nauseated and reduce their eating, particularly of sweet, high-fat foods,[29] but would not alter consumption of regular foods, such as protein, fruits and vegetables. Naltrexone (Revia) is a cheap, safe compound that can damp down the reward system of drug addicts but does relatively little in everyone else. A subset of our SHINE subjects took one naltrexone pill to acutely block the reward system. We noted that those who scored highest on the RED scale at baseline were the ones who developed nausea in response to the naltrexone, suggesting that these subjects had the highest baseline EOP function.[30] In another study, we found that obese women highest on the RED scale had the largest reductions in food-craving intensity after taking the naltrexone pill.[31] In these series of studies, we found that the people with the highest reward-driven behavior were the most responsive to our blocking of their reward system. Therefore, we were pretty sure that we had found a probe to study the biology of food addiction.

Lastly, we needed to find out whether we could affect this system with a behavioral intervention (e.g., mindfulness) that reduces food cravings and stress. We followed these subjects through their diet and exercise interventions either with or without the mindfulness intervention. Perhaps it shouldn't be surprising that those who received the mindfulness eating and stress intervention, rather than the more general intervention that didn't focus on reward-driven eating, experienced large reductions in their reward-driven eating, and that these reductions in reward-driven eating led to weight loss.[32] What's more, it was really about the reductions in reward-related eating—not the reductions in stress—that were key to weight loss. What was interesting was that these vulnerable people, with their broken reward systems, showed these improvements more so in the meditation condition, when they got the extra training to improve their mindfulness (see Chapter 18).

In this series of experiments, we showed that:

- Not everyone with obesity is the same.
- Some people experience a loss of control with certain foods.
- Those that do tend to binge on high-sugar/high-fat foods (think chocolate cake).
- This aberrant eating behavior is driven by dysfunction of the reward system.
- The stress system piles on to disinhibit cognitive control of food intake.
- Blockading the reward system unmasks both the reward and stress systems.
- Mindfulness can restore functionality to the reward and stress systems, leading to improved mood, less disordered eating, and less risk for metabolic syndrome.

So here's the question. Let's say you're one of these sugar-addicted people. Maybe you employ lots of restraint to stay away from the obvious triggers: soda, cakes, ice cream. But you still have to eat. And what if your food has sugar mixed or baked right into it and you don't know it? Can you break an addiction if the addictive substance is so pervasive that it's in everything but you don't know it's there?

Then on top of the added sugar, go ahead and reduce the other two molecules from Chapter 9 that increase contentment: tryptophan and omega-3 fatty acids. Tryptophan is very low in processed food because protein sources of tryptophan are relatively expensive. Omega-3 fatty acids are even more expensive, and tend to provide a fishy odor to food. Processed food is high sugar, low tryptophan, low omega-3s. Great for reward but risky for both addiction and depression.

Winning the Battle Against Big Sugar

This, my friends, is the explanation for America's, and the world's, love affair with processed food. By slowly adding sugar not just to desserts but to diet staples and condiments as well, the food industry has been able to hook us and keep us hooked (see Chapter 14). The fat and the salt, while not addictive themselves, serve to increase the salience of the added sugar (see Chapter 6).[33] Then, of course, add the second legal addiction—caffeine—to soft drinks, energy drinks, coffee beverages, and the like to provide the second hook. The bitterness of the caffeine is more than offset by the sugar. Plus caffeine increases the salience, or rewarding properties, of sugar.[34] Two addictions—and it's all completely legal, because both sugar and caffeine are Generally Recognized as Safe (GRAS) by the FDA. That means the processed food industry is allowed to use any amount it wants in any food it chooses, and with no repercussions. This designation of GRAS is the least policed administrative law in all of Washington. The GRAS determination provides the underpinning for the success of the entire processed food industry. This is how and why the processed food industry changed our food supply forty years ago, and why we've gotten sick and unhappy ever since. And it's why every country that has adopted our food supply has suffered the same fate. Even the U.S. Government Accountability Office says that GRAS is dangerous as a designation,[35] and lawyers are starting to call for the revocation of sugar's GRAS status,[36] similar to how trans fats were removed from the GRAS list. A group of doctors, lawyers, and entrepreneurs in San Francisco have started a nonprofit organization called EatREAL (eatreal .org) to reverse diet-related disease by changing the global food supply. One of our long-term goals is to get sugar removed from the FDA's GRAS list.[37] I am proud to serve as chief science officer.

The good news is that after decades of tallying record profits right up until 2012, the soft-drink and fast-food industries all of a sudden aren't doing so well. McDonald's, Coca-Cola, and Pepsi consistently outperformed the S&P 500 for the previous three decades. In 2014, Muhtar Kent, then CEO of Coke, with declining profits, announced the firing of eighteen thousand employees (although the savings was going to be plowed back into advertising, especially to children), and in 2015 Don Thompson, then CEO of McDonald's, was fired for poor performance. British sugar company Tate & Lyle lowered expectations on 2015 profits due to declining sugar demand. Some companies have recognized the problem and are trying to get ahead and even take advantage of this global trend of sugar reduction in processed food. For instance, in the Netherlands, the grocery chain Albert Heijn has pledged to reduce the added sugar in hundreds of their store-brand grocery items, including yogurt, cookies, custard, and ketchup.[38]

Furthermore, in response to the global obesity and diabetes epidemics, several countries have examined the research themselves. They are working to oppose the entrenched food industry lobbyists, and some have enacted a sugar excise tax (soda and junk food) in order to get people to reduce consumption by reducing effective availability. Thus far, Mexico and the UK have enacted such a tax,[39] while Australia, New Zealand, South Africa, India, and even Saudi Arabia are considering similar legislation. Closer to home, we've seen six American cities—San Francisco, Oakland, Berkeley, and Albany, California, as well as Chicago and Philadelphia—enact sugar taxes to generate money for programs and reduce consumption. No information on changes in health or happiness from any of these governmental maneuvers as of yet. But, not surprisingly, there is quite a bit of pushback from both the lobbyists and the people who feel this falls strictly within the realm of personal responsibility. But does it?

In other countries around the world, sugar consumption within soft drinks and processed foods is on the rise[40] because of lax rules and, just like what happened to tobacco: once America started tightening the screws on cigarette availability and use, the tobacco companies moved offshore in an effort to addict new populations. And in these countries the sugared-beverage industries have a leg up, because people have to drink but they don't trust their water supplies.[41] And who provides the water-purification apparati for most third-world nations? You guessed it . . . Coca-Cola.

Sugar "Pop-Aganda"

The food industry has kept us off balance for years by deflecting criticism with its "commonsense" yet fallacious refrain, "A calorie is a calorie." It's the quantity of those calories, not the quality. This has been the core of its business strategy for at least fifty years. My public policy colleagues at UCSF have unearthed industry documents dating back to 1965 that show that the Sugar Research Foundation (the PR arm of the sugar industry) paid two scientists a handsome sum ($50,000 in today's dollars) to publish two reviews in the *New England Journal of Medicine* that exonerated sugar and pinned the blame for heart disease on saturated fat.[42] Furthermore, they also showed that in 1971 the corporate suspects infiltrated the NIH agenda to steer research on dental cavities away from sugar reduction and instead to promote a vaccine that never materialized.[43] The food industry continues to put its thumb on the scales of objectivity, both figuratively and literally. Five out of six studies funded by the industry show no effect of sugared beverages on weight gain, while ten out of twelve studies by independent scientists show a clear effect on weight gain.[44] More recently, Coca-Cola was exposed in paying off three scientists to form the Global Energy Balance Network to pin the blame for the obesity epidemic on lack of exercise.[45] In fact, the soda industry has given away a total of more than $120 million to ninety-six separate public health organizations to promote anything but food industry regulation.[46]

The processed-food industry has another trick up its sleeve as well: redefining portion size. Consider processed peanut butter. If you spread a bagel with peanut butter, a standard serving is two tablespoons containing 188 calories, with 3 grams of sugar. Very few people use only two tablespoons for a PB&J. Yet Nutella, in a battle for stomach-share of peanut butter-gorging kids everywhere, and in an attempt to undercut its competition, successfully argued that it should be reclassified as a "jam" rather than as a peanut butter, and so its portion size should read lower, at only one tablespoon, with only 100 calories (half as much as their competitor). Nutella is hoping you'll miss the 10.5 grams of added sugar.[47]

Worse yet, our processed food diet has been engineered by the food industry to be "fortified" with all sorts of vitamins, minerals, and a panoply of additives—whatever's trending (lycopene, flavonoids, resveratrol)—supposedly to give us all the nutrients we need. This is the basis of the $121 billion nutraceutical industry. You need great hair! Great skin! Perky breasts! And the "natural" key is: bull semen, jojoba, lychee, raspberry ketone, açaí berry. You should cleanse, get an enema, juice! There's always a new pop-up clickbait on your internet news feed with a miracle cure for being happy and never aging. The FDA doesn't regulate nutraceuticals and companies don't have to show efficacy: after all, they're "food." But you'll never see "What I Did to Be Happy!" clickbait articles for the brain biochemical that can lead to happiness—the right kind of real food.

At the other end of the spectrum, the diet sweetener industry has argued that "a calorie is a calorie" means that their products, by providing sweetness without calories, is the better choice and certainly the best choice for those with obesity. Artificially sweetened beverages now account for one-quarter of the global market. Except for one little problem: they don't work. A recent meta-analysis of all the studies substituting artificially sweetened beverages for the full-calorie version showed absolutely no change in weight and displayed a now-familiar bias: those funded by industry show weight loss, while those conducted by independent scientists show absolutely no change in weight.[48]

Processed Food: An Experiment That Failed

The science is in. Processed food is addictive, can make you extremely unhappy, and may ultimately kill you.[49] Processed food is the exposure you can't escape, because it's everywhere. Except you can—but you have to be mindful of it in order to do it. Washington will never get on board, as protection of American business trumps protection of the American public. So it's up to each and every one of you. Based on the science presented throughout this book, I offer to you my single most important key to happiness: **COOK *REAL FOOD* FOR YOURSELF, FOR YOUR FRIENDS, AND FOR YOUR FAMILY!** It's *connection* in that you will be sitting down with people you like (and maybe even love); it's *contribution* because you are making something worthwhile; it's focusing so it's easier to *cope*; and, unless you spike it with something, it's *non-addictive*. The amount of sugar in processed food far exceeds what you would include yourself. If you use real ingredients, it will be delicious. It's one of the key ingredients to contentment. And real food means low sugar and high fiber; the fiber feeds your microbiome so your bacteria will be happy as well. You may lose weight, and you will definitely reduce your risk for all of the chronic diseases of metabolic syndrome. And you will be sticking it to the companies who are trying to addict you and your family.

The problem is that one-third of Americans currently don't know how to cook; they've fallen prey to the food industry's endgame. Microwaving is not cooking: it's boiling water. If you don't know how to cook, you're hostage to the food industry for the rest of your life and unwittingly will pass this on to your children. You can farm out the shopping; there are companies that will buy the real food and deliver it to your door for you to assemble. Similarly, you can farm out the cooking: there are other companies that will dirty their kitchen instead of yours. Getting ahead in life isn't just taking extracurricular classes and joining the debate team—it's actually cooking and (shocker) spending time with each other. Not just at the table looking down at your favorite gadget, but actually engaging in meaningful conversation. There is nothing that will improve your health, your well-being, your achievement, your sense of accomplishment, your sense of community, and the health and happiness of your family as much as cooking for yourself and enjoying a meal with others. It costs time, to be sure. But it saves money—lots of money—both in food costs and in medical bills.

Processed food is no different from any other substance of abuse. Technology brings along its attendant multitasking and sedentary behavior and sleep deprivation. You may have fallen prey to any or all of these ploys over the last forty years. Yet you didn't succumb to any of these rewards because you needed them; rather, you wanted them. Maybe because everyone else wanted them, too, as conformity is its own form of stress. Nonetheless, you bought your way in. Just like they planned.

Your Mother Should Know

America is home to the corporate consumption complex, but this problem exists all over the world. We stopped being individuals decades ago after the advent of GDP; we're all consumers now. Technology, sleep deprivation, substance abuse, processed food—these are the killers of contentment and the drivers of desire, dependence, and depression. *Connect, contribute, cope, cook*: each of these has the capacity to pull you out of addiction by limiting the need for reward by optimizing the effects of dopamine and reducing cortisol—and lift you out of depression by increasing contentment and the effects of serotonin. None of these strategies are new, although the science behind them is. These are all things your mother told you when you were a kid, when you were still growing, but since then you may not have had time to adopt or enact any of them, because you have been too busy texting while quaffing a Coke—because the siren call of pleasure is just too great, and too immediate.

Happiness, our first garden, is our natural birthright, but we've been cheated out of it. In its place has been substituted a garden of earthly delights, and we're all the worse for it. Some pay the ultimate price and slip into the abyss of eternal damnation. But that first garden is right in front of you, just behind the curtain of your own brain. You can reenter anytime you choose. I've chosen. I suggest you choose now. It's time to reclaim your original garden as your own.

Epilogue

n 2014, I visited a major American medical school to give Psychiatry Grand Rounds on sugar and addiction. The administrator of the hospital's substance abuse recovery program was a woman in her late forties, a former drug addict who had pulled herself out of her misery from opiate use. On being asked what addiction and getting clean meant to her, she replied, "When I was shooting up, I was happy. What my new life has brought me is pleasure." When I heard this, I was quite taken aback. Of course, it's exactly the opposite. People shoot up to recapture the pleasure of their very first hit. But they never can. So they inject more and more to derive less and less pleasure. It is not a coincidence that this woman misconstrued pleasure for happiness; this is exactly why she was an addict, albeit now in recovery.

I have known about the dopamine-serotonin-cortisol connection for at least three decades, dating back to my postdoctoral fellowship in a neurobiology lab at the Rockefeller University in New York. But it wasn't until meeting and talking to this semi-unfortunate woman that I recognized how this seemingly trivial confusion might figure prominently in terms of why people become addicted in the first place. If she was so sure that she was happy while shooting up, I figured others might feel the same way. I've since talked to many of my colleagues in psychiatry and substance abuse treatment, and they corroborate that this view is commonly held among their patients.

Shortly thereafter, I was in Minnesota on a family vacation. My sister-in-law used to run the consumer response department at Pillsbury in Minneapolis before they were bought out by General Mills in 2001. She had to deal with all the phone calls from irate customers when the Poppin' Fresh dough didn't rise or when there was freezer burn on the ready-to-bake biscuits or crescent rolls. Although she has been gone from that job for over a decade, she is still convivial with the crew she worked with, and they see each other once a year or so for a gourmet club. One of her friends who had undergone bariatric surgery several years prior commented to my sister-in-law, "You look wonderful! So nice and slim. How do you do it?" She said, "I don't need to eat much. I don't eat when I'm not hungry." To which her friend responded, "Don't eat? Who eats for hunger? We're not hungry either. Eating is about happiness."

That was the aha moment. How many other people get it wrong? Of course, I take care of obese children on a daily basis, and I get to see and hear what their parents feed them and what they eat when they're on their own. I knew this issue of eating for happiness wasn't just anecdotal. Many patients tell me "food is my friend"—after all, it's always there when they need it, it comforts them, and it never leaves their side. Well, you could argue, that's exactly what's wrong with people with obesity: they eat when they're not hungry! And for a certain segment of obese people, there's a lot of truth in that statement. But that leaves two questions that demand answers. First, why do they eat for happiness and how did they get that way? What made them need a surrogate friend, one that doesn't talk back? And second, there are a whole lot of thin people who also eat when they're not hungry, and they manifest the same diseases as do the obese, such as type 2 diabetes, heart disease, fatty liver disease, hypertension, cancer, and dementia; all of the diseases of metabolic syndrome. How many thin people are destined to succumb to one or more of these diseases, just because they didn't know?

Either of these two clinical vignettes might have been passed off by the uninitiated as purely anecdotal. But the science and my clinical experience said otherwise. And as I researched the data for this book, the dark underbelly of Western culture and how it manipulates our beliefs and behaviors became ever more painfully exposed. We take it for granted that our society values money and its pleasures over all else, and then conflates those emotions with happiness. If parents don't teach it directly to their children, then the TV or the internet will. This book just had to be written.

How many people are addicted to either a behavior or a substance, and they think it's just a part of their general personality? They might say, "Oh, I have a horrible sweet tooth," or "I'm a chocoholic from way back," or that they frequently engage in retail therapy and post about it on Facebook for validation. No one comes out of the womb that way. You have to activate the dopamine pathway first. It's reward, but it's also learning—"This feels way good." Once exposed, each of these behaviors becomes reinforced through activation of the reward pathway. And then the receptors start to dwindle. In no time, each individual becomes just another member of the mainstream consumer culture, another cog in the wheel of our economy, which boasts hedonic substances as commodity numbers two (coffee), four (sugar), and eight (corn, which is turned into high-fructose corn syrup).

I hope this book conveys that there's nothing inherently wrong with pleasure—but not to the exclusion of happiness. Pleasure and happiness are not mutually exclusive, although in this book I've separated them as much as possible so as not to confuse the reader. After all, Wall Street, Las Vegas, Madison Avenue, Silicon Valley, and Washington, D.C., have confused enough people. My goal for this book was to scientifically parse the difference between these two ostensibly positive emotions, examine them separately, and watch what happens when you recombine them. You can, and should, have *both* pleasure *and* happiness in your lives. One will make the other that much sweeter. There are moments in life when you can experience both simultaneously (your team winning the Super Bowl or an Olympic gold medal, attending a wedding, going on a great vacation, having a child, or finishing a job well done at work), which elevates the baseline feeling of contentment to joy or elation, and we might find those experiences nothing short of rapturous. Sometimes the amplitude of simultaneous emotion is so great that we cry. These events are likely to leave the largest imprint on our memories, and will likely stay with us for the rest of our lives. In the future, when we pull them out of our subconscious and examine them, the sense of reward will have long dissipated, but the contentment within the memory will still be there. And don't forget, things that generate pleasure often can be expensive, but things that generate happiness are dirt cheap.

The prediction of our demise due to our quest for pleasure is attributed to Aldous Huxley, who pronounced, "What we love will ruin us." In *Brave New World* (1932), he described a human race that by the year 2540 had been destroyed by ignorance, technology, constant entertainment, and material possessions. But his forecasting was off by four centuries, as we're already there. Conversely, Tolstoy buffs will recall that in *War and Peace* (1865), after protagonist Pierre Bezukhov is incarcerated by the French, he has ample time to ponder the meaning of existence. Awed by the serenity of a fellow prisoner, Pierre learns "not with his intellect but with his whole being . . . that man is created for happiness, that happiness is within him, in the satisfaction of simple human needs, and that all unhappiness arises not from privation but from superfluity." In his captivity, Pierre postulates the virtuous simple life as the shortest route to contentment: "The satisfaction of one's needs—good food, cleanliness, and freedom—now that he was deprived of all this, seemed to Pierre to constitute perfect happiness."

The keys to benefit from pleasure and happiness are to understand the differences between the two, because even though pleasure and happiness are not mutually exclusive, they can still be opposites. There is plenty of room for pleasure in life, and lots of *things* can bring you pleasure. But no *thing* can make you happy. *Experiences* can make you happy. *People* can make you happy. *You* can make you happy. There are many ways to get there, and I've outlined them in this book. Each of them necessitates that you peel back the curtain of your own brain. There are many obstacles—your boss, your friends, your family, and of course even you—and they will derail you, but only if you let them, and only if you don't know the difference. To paraphrase Benjamin Franklin, a great pleasure seeker himself: Those who abdicate happiness for pleasure will end up with neither. The science says so.

ACKNOWLEDGMENTS

As stated in the introduction, I am not a psychiatrist or an addiction physician. But I am a quick study. And I've had some very good teachers along the way.

I work with five different teams of scientific investigators and clinicians, each of which has helped me to understand the concepts that pervade this book. First, my colleagues from the UCSF Osher Center for Complementary Medicine have been instrumental in helping to establish the concept of reward-eating drive and the addictive effects of sugar. Elissa Epel has been my friend since the day I arrived at UCSF. The rest of the Osher group, including Ashley Mason, Nicole Bush, Jeff Milush, Eli Puterman, Doug Nixon, Patty Moran, Kim Phox-Coleman, Jennifer Daubenmier, Barbara Laraia, Mary Dallman, Peter Bacchetti, Michael Acree, and Fredrick Hecht, have worked tirelessly to benefit obese patients. Second is my metabolic team from both UCSF and Touro University. Together we have uncovered the toxicity of sugar and its role in fomenting chronic metabolic diseases by removing it from childrens' diets. I am indebted to Jean-Marc Schwarz, Kathy Mulligan, Alejandro Gugliucci, Susan Noworolski, Viva Tai, Mike Wen, and Ayca Erkin-Cakmak. Third are my colleagues at the UCSF Institute for Health Policy Studies, who recognize that non-communicable diseases are the greatest global health threat in the history of mankind—and they're all due to hedonic substances. First it was alcohol, then it was tobacco, and now it's sugar. Laura Schmidt, Stan Glantz, Claire Brindis, and Cristin Kearns understand the intersection of science and policy, and have helped translate the work into meaningful public health messaging. Fourth, my colleagues at the UCSF Center for Global Health, who can take policy and model it into money (saved), because this is the only thing that governments care about. Jim Kahn, Rick Vreman, Alex Goodell, and my dietitian and current grad student Luis Rodriguez have all been instrumental in expanding the message beyond the American border. And finally my clinical team at the UCSF Weight Assessment for Teen and Child Health (WATCH) clinic, where instead of treating the obesity (which doesn't work) we treat the metabolic dysfunction (the insulin problem) and watch the pounds melt away. Patrika Tsai, Kathryn Smith, Meredith Russell, Luis (again), Nancy Matthiesen, and Megan Murphy are the reason it's still fun to go to clinic each day. I also must thank Marc and Lynne Benioff for believing in this work and for their support of our research at UCSF.

I also work with many psychiatrists, some of whom double as rat researchers studying dopamine and serotonin, and who have helped advance the concepts in this book. Among them are Larry Tecott, Steve Bonasera, Nicole Avena, Ashley Gearhardt, Mark Gold, Rajita Sinha, Ania Jastreboff, Mark Potenza, Eric Stice, Mark George, and Jeff Kahn. I've also had some very informative discussions with Steve Ross, Brian Earp, Adam Gazzaley, and John Coates during a fun weekend in Mountain View, and they helped to fill in some of the pieces. One person who deserves special mention is Bill Wilson (not that one), a tremendously astute clinician who can see patterns in patients, and who helped me understand the role that carbohydrates play in brain dysfunction, confusion, and unhappiness.

The arc of this book also dovetails into history and law. I particularly want to thank my law mentors turned colleagues David Faigman and Marsha Cohen of UC Hastings College of the Law. Also a nod to Michael Roberts of the UCLA Resnick Center for Food Policy and Law.

Here in the U.S., we started a non-profit called the Institute for Responsible Nutrition, now the science arm of a new entity called EatREAL, to provide the science to alter the global food supply, in order to eradicate childhood obesity and type 2 diabetes. My colleagues Jordan Shlain, Wolfram Alderson, Lawrence Williams, and our supportive board have been instrumental to furthering the cause. We also work with two sister organizations here in San Francisco: Wellness City Challenge with Cindy Gershen and Pam Singh, and Andrea Bloom of ConnectWell. Julie Kaufmann of the American Heart Association has also been an enthusiastic supporter.

The battle for our food supply is an international one, and I've got some amazing colleagues stretched over several continents. In the UK, Aseem Malhotra and David Haslam of the National Obesity Forum spelled out the danger, Jack Winkler helped craft the argument, and Graham MacGregor and colleagues at Action on Sugar had the government's ear as the UK Sugar Tax was enacted. In Mexico, I am indebted to Juan Rivera Dommarco of the Instituto Nacional de Salud Publica and Juan Lozano Tovar of the Ministry of Social Security of Mexico for their work behind the scenes on the Mexico Soda Tax and benefiting the Mexican people. In the Netherlands, Albert van de Velde, Martijn van Beek, Peter Klosse, Hanno Pils, Peter Voshol, and Barbara Kerstens of the non-profit Voeding Leeft (Food Lives) are fighting the good fight. And in Australia and New Zealand, my colleagues Rory Robertson, Gary Fettke, Simon Thornley, Kieron Rooney, Rod Tayler, David Gillespie, Sarah Wilson, and Gerhard Sundborn have made a remarkable stand against entrenched forces in both academe and parliament that seek to undermine the population Down Under. I'm also indebted to some unlikely international allies: members of the finance profession. Stefano Natella, formerly of Credit Suisse, is a true Renaissance man. No one gets the problem better than he does, and he is using his position to sway business to forge a better society. Caroline Levy and Philip Whalley of CLSA also deserve a great deal of credit for helping to move the message about the role of sugar in the food supply.

Other academic faculty have contributed greatly to my knowledge. My UCSF ex-boss Walter Miller has always been a role model of the uncompromising scientist. Synthia Mellon and Owen Wolkowitz taught me about steroids and brain function. Nobel Laureate Stan Prusiner and my best friend in science Howard Federoff have helped to delineate the role of sugar as a potential driver of dementia through fatty acids in the brain. Dieter Meyerhoff explained the neuroimaging of alcoholism. Justin White tutored me in behavioral economics. Michele Mietus-Snyder counseled me on lipids and mitochondria; Laurel Mellin on stress, negative emotion, and its effects on obesity; Nancy Adler on vulnerable populations and social determinants of disease; Jyu-Lin Chen on East Asian populations; and Alka Kanaya on South Asians. I also want to thank Alexandra Richardson of Oxford University, who taught me the magic of omega-3s; Dacher Keltner, who teaches a course in positive psychology at UC Berkeley; and John Graham, a professor of marketing at UC Irvine.

Others have helped to espouse the public health message. Kelly Brownell is the dean of the Sanford School of Public Policy at Duke University. Matt Richtel is a Pulitzer Prize–winning author who has written about the texting and cell phone issue. Jim Steyer of Common Sense Media was the first to Talk Back to Facebook. Many people in the media have also been integral to dissemination of the message. Corby Kummer of the *Atlantic*, Patty James and Bill Grant of the Commonwealth Club, Laurie David and Stephanie Soechtig of *Fed Up* (2014), Michele Hozer and Janice Dawe of *Sugar Coated* (2015), and author Gary Taubes, author of *The Case Against Sugar* (2017), are all allies.

One weekend in August 2015, my friends Fred Aslan and Jack Glaser (as well as Jack's wife, Elissa Epel, and her parents) and I unwittingly spent a silent meditation weekend in the mountains with Buddhism instructor James Baraz. Fred, Jack, and I all deserve gold stars for making it through that weekend, and James as well for putting up with the three of us. You try it sometime. We were like kids in detention. But that experience was seminal to integrate the religion together with the practice of mindfulness together with the neuroscience.

This book was particularly difficult to write, due both to its tenor and its content. My agent Janis Donnaud and publisher Caroline Sutton had to deal with several work stoppages due to grants, personal illness, and the death of my mother. They could have pulled the plug, but they didn't, because they believed. I'm indebted to them for seeing it through to the end. I also need to thank my personal friends Matt Chamberlain, who kept my internet up and running, and my graphics design gurus Glenn Randle and Jeannie Choi. Mark Gold, Walt Miller, Elissa Epel, David Faigman, Ashley Mason, Bill Wilson, and Kathy Laderman read early versions of the manuscript and provided comments.

My penultimate thanks go to my writing team: my editor Amy Dietz and my researcher Deanna Wallace. Amy's father was a neurologist who named her for the amygdala, a bit of foreshadowing of her life to come. This is our second foray together, and we haven't killed each other yet. Amy brought three special qualities to this book. First, she obtained her master's in public health at the University of Washington and is able to look at health problems through the wide lens and break down difficult concepts for the lay reader. As someone in recovery, she was able to bring her own experiences and those of her friends to this book, which both humanized the often distasteful topic of addiction and also helped me to understand what I was actually writing about. And lastly she is a hoot. This book could have been so dry it would crumble, but Amy's wit and facility with popular culture has made it juicy and relevant, and hopefully gave you a chuckle. Deanna has broad expertise in reward behavior, dopamine, and cognitive neuroscience, obtaining her Ph.D. with Eric Nestler at the University of Texas Southwestern Medical Center, completing postdoctoral training with Mark D'Esposito at UC Berkeley, and currently working as a researcher at UC San Francisco. She is also married to Mike Donovan, a serotonin researcher at UCSF.

Deanna was the "bulls—t detector" of this project. She vetted the scientific articles to make sure they said what I thought they did and made sure they passed muster, sometimes to my chagrin. She was a gifted backstop and is a truly remarkable human being. Yet Amy and Deanna are completely different: Amy is gregarious and Deanna demure. On different days, my ranting and raving would invariably put one of them up a wall. But I couldn't have written this book without them, and I love them both.

Lastly, my most sincere thanks go to my family. I am indebted to my sister Carole Lustig-Berez for handling my mother's affairs at a most difficult time, and for being our family's primary support at a time when I couldn't. Also thanks to my brother-in-law Mark Berez for funneling me media reports on nutrition and the business backlash. And I must reserve my most grateful appreciation to my wife, Julie, and my daughters Miriam and Meredith, whom I love more than anything. Writing this book put a strain on all of us for almost two years. Between work, travel, and book writing, I sometimes wasn't there for family events that I should have been there for. And I heard about it. The good news is we bent but didn't break. And now that this book is "in the can," I can go back to being a husband and father, the two things that make me happy.

GLOSSARY

Addiction: a strong and harmful need to regularly have something (such as a drug) or engage in a specific behavior (such as gambling) due to an overwhelming biochemical drive, and which cannot be controlled by behavioral restraint.

ALT: alanine aminotransferase, a blood test that tells about liver function and is very sensitive for the amount of fat in the liver.

Amygdala: part of the stress-fear-memory pathway. This walnut-sized area of the brain generates the feelings of fear and stress, which tells the hypothalamus to tell the adrenal glands to make extra cortisol.

Anandamide: a naturally occurring neurotransmitter that binds to the CB1 endocannabinoid receptor and that reduces levels of anxiety.

Apoptosis: programmed cell death, in which proteins in the cell are activated to cause self-destruction.

Autonomic nervous system: that part of the nervous system that controls unconscious functions of the body. It consists of two parts: the sympathetic system controls heart rate, blood pressure, and temperature; while the parasympathetic system (the vagus nerve) controls eating, digestion, and absorption, slows the heart rate, and lowers blood pressure. The two together control energy balance.

Cortisol: the stress hormone released from the adrenal glands, which acutely mobilizes sugar for use but which chronically lays down visceral fat and also reduces serotonin-1a receptor number.

Depression: a mental condition characterized by feelings of severe despondency and dejection, inadequacy, and guilt, often accompanied by lack of energy and disturbance of appetite and sleep.

Developmental programming: alterations in brain or body functioning due to alterations in the environment that occur in the fetus prior to birth.

Dopamine: part of the reward pathway. A neurotransmitter that, when released, can acutely cause feelings of reward but, when released, chronically reduces the number of its receptors, leading to tolerance.

Dopamine receptor: the protein that binds dopamine to generate the reward signal and, when reduced in number, leads to tolerance.

Endocannabinoid: a neurotransmitter, such as anandamide, that binds to brain receptors and acts like marijuana, driving reward and reducing anxiety.

Endogenous opioid peptide (EOP): part of the reward pathway. A neurotransmitter made in the brain that binds to its receptor to signal the consummation of reward or euphoria.

Endogenous opioid peptide (EOP) receptor: part of the reward pathway. A protein that binds either opiates (e.g., heroin) or endogenous opioid peptides (e.g., beta-endorphin) to signal the consummation of reward or euphoria.

Epigenetics: modifications in DNA without changes in the DNA genetic sequence, usually occurring prior to birth.

Estrogen: female sex hormone, made either in the ovary or in fat tissue.

Excitotoxicity: the process of overstimulating a cell, leading to cell dysfunction or death.

Fructose: half of dietary sugar or high-fructose corn syrup. The molecule that makes sugar taste sweet causes the reward system to activate, and is the addictive component.

Glucose: half of dietary sugar or high-fructose corn syrup. Also the molecule found in starch, the molecule that every cell on the planet burns to liberate energy.

Hippocampus: part of the stress-fear-memory pathway. The part of the brain where memories are housed and that exerts influences on the amygdala and prefrontal cortex.

Hypothalamus: part of the stress-fear-memory pathway. The area at the base of the brain that controls various hormone systems of the body, particularly cortisol.

Insulin resistance: the state where insulin signaling is reduced, requiring the beta-cells of the pancreas to make more insulin, which drives both obesity and chronic disease.

Insulin secretion: the process of insulin release in response to both rising blood glucose and the firing of the vagus nerve.

Insulin: a hormone made in the pancreas that tells fat cells to store energy and interferes with the leptin signal to increase food intake.

Leptin resistance: the state where the leptin signal is dampened, leading to the hypothalamus interpreting starvation.

Leptin: a hormone released from fat cells that travels in the bloodstream to the hypothalamus to report on peripheral energy stores.

Major depressive disorder (MDD): a mental disorder characterized by at least two weeks of low mood, self-esteem, loss of interest in normally enjoyable activities, low energy, and pain without a clear cause, often needing medical treatment.

Metabolic syndrome: a cluster of chronic metabolic diseases characterized by energy overload of cells.

Micronutrient: vitamins or minerals found in real food, usually isolated with the fiber fraction.

Mitochondria: subcellular organelles specialized to burn either fat or carbohydrate for energy.

Neurotransmitter: a chemical in the brain made in one nerve cell, which, when released, causes other nerve cells to fire.

Necrosis: cell death due to exposure to a toxin or lack of blood or oxygen.

Nucleus accumbens (NA): the area of the brain that receives the dopamine signal and interprets the feeling as reward.

Obesity: excess body fat deposition.

Omega-3 fatty acids: a fatty acid found in wild fish and flax that is an important component of neuronal membranes and that reduces inflammation.

Phenylalanine: a dietary amino acid that can be converted into dopamine.

Prefrontal cortex (PFC): part of the stress-fear-memory pathway. The part of the brain, located in the front (above the eyes), that inhibits impulsive and socially unacceptable and potentially dangerous behaviors and actions.

Peptide YY(3-36): a hormone made by the small intestine in response to food that signals satiety to the hypothalamus.

Satiety: the feeling of fullness that stops further eating.

Serotonin: part of the contentment pathway. A neurotransmitter that, when it binds to its -1a receptor on neurons, transmits feelings of contentment; and, when it binds to its -2a receptor, evokes the "mystical experience."

Serotonin-1a receptor: part of the contentment pathway. A protein on the surface of neurons that, when it binds serotonin, reduces neurotransmission, which leads to feelings of contentment.

Serotonin-2a receptor: a protein on the surface of neurons that, when it binds serotonin, evokes the "mystical experience."

Stress: an uncomfortable state of mental or emotional strain or tension resulting from adverse or demanding circumstances. Accompanied by neural output from the amygdala, which tells the hypothalamus to signal the adrenal glands to make the hormone cortisol.

Subcutaneous fat: the fat outside of the abdomen, which is a storehouse of extra energy but which does not signify an increased risk for metabolic syndrome.

Sympathetic nervous system: the part of the autonomic nervous system that raises heart rate, increases blood pressure, and burns energy.

Tetrahydrocannabinol: the active substance in marijuana that binds to the CB1 endocannabinoid receptor to reduce levels of anxiety.

Tolerance: the state where the signal for reward is dampened and can only be generated by consuming more substrate (e.g., sugar) or engaging in more behaviors (e.g., gambling).

Transcription factor: a protein in cells that turns on genes to make the cell change its function.

Tryptophan: a dietary amino acid that is converted into serotonin.

Type 2 diabetes: a disease of high blood sugar due to defective insulin action on tissues.

Tyrosine: an amino acid (which can be consumed, or derived from phenylalanine) that can be converted into dopamine.

Vagus nerve: part of the autonomic nervous system that promotes food digestion, absorption, and energy storage, and slows heart rate.

Ventral tegmental area (VTA): part of the reward pathway. The area of the brain that sends the dopamine signal of signifying reward to the nucleus accumbens.

Visceral fat: the fat around the organs in the abdomen, which is a risk factor for diabetes, heart disease, and stroke, and a marker for metabolic syndrome.

NOTES

CHAPTER 1. THE GARDEN OF EARTHLY DELIGHTS

1. McMahon DM, *Happiness: A History*. Grove, New York (2006).
2. Oishi S et al., "Concepts of Happiness Across Time and Cultures." *Personal. and Soc. Psychol. Bull.* 39, 559–77 (2013).
3. McMahon DM, *Happiness: A History*. Grove, New York (2006).
4. Helliwell J et al., World Happiness Report (2015). http://worldhappiness.report /wp-content/uploads/sites/2/2015/04/whr15.pdf
5. Oxford Happiness Questionnaire. http://interactive.guim.co.uk/embed/labs/2014/dec /01/happiness/oxford-happiness-quiz-questionnaire/index.html
6. Boehm JK et al. in *Handbook of Positive Psychology*, Lopez SJ, ed., Oxford University Press, Oxford (2015).
7. Bartels M, "Genetics of Wellbeing and Its Components Satisfaction with Life, Happiness, and Quality of Life: A Review and Meta-Analysis of Heritability Studies." *Behav. Genetics* 45, 137–56 (2015).
8. Okbay A et al., "Genetic Variants Associated with Subjective Well-being, Depressive Symptoms, and Neuroticism Identified Through Genome-Wide Analyses." *Nat. Genet.* 48, 624–33 (2016).
9. Kharpal A, "A Scientist Has Discovered Why Happiness Might Very Well Be Genetic." CNBC, Feb. 11, 2017. http://www.cnbc.com/2017/02/11/a-scientist-has-discovered-why -happiness-might-very-well-be-genetic.html
10. *Stanford Encyclopedia of Philosophy.* http://plato.stanford.edu/entries/happiness
11. Freudenberg N, *Lethal but Legal*. Oxford University Press, New York (2014).

CHAPTER 2. LOOKING FOR LOVE IN ALL THE WRONG PLACES

1. Freudenberg N, *Lethal but Legal*. Oxford University Press, New York (2014).
2. Bradford AC et al., "Medical Marijuana Laws Reduce Prescription Medication Use in Medicare Part D." *Health Aff.* 35, 1230–6 (2016).
3. De Boer A et al., "Love Is More Than Just a Kiss: A Neurobiological Perspective on Love and Affection." *Neuroscience* 201, 114–24 (2012).
4. Coria-Avila GA et al., "Neurobiology of Social Attachments." *Neurosci. Behav.* Rev. 43, 173–82 (2014).
5. Pedersen CA et al., "Maternal Behavior Deficits in Nulliparous Oxytocin Knockout Mice." *Genes Brain Behav.* 5, 274–81 (2008).
6. Brewerton TD, "Hyperreligiosity in Psychotic Disorders." *J. Nerv. Ment. Dis.* 182, 302–4 (1994).
7. Takahashi K et al., "Imaging the Passionate Stage of Romantic Love by Dopamine Dynamics." *Front. Hum. Neurosci.* 9, 191 (2015).
8. De Boer A et al., "Love Is More Than Just a Kiss: A Neurobiological Perspective on Love and Affection." *Neuroscience* 201, 114–24 (2012).
9. Siegler IC et al., "Consistency and Timing of Marital Transitions and Survival During Midlife: The Role of Personality and Health Risk Behaviors." *Ann. Behav. Med.* 45, 338–47 (2013).
10. Palmer R, "Addicted to Love" (1985). https://www.youtube.com/watch?v=XcATvu5f9vE.
11. Earp BD et al., "Addicted to Love: What Is Love Addiction and When Should It Be Treated?" *Philos. Psychiatr. Psychol.* (2015). http://www.academia.edu/3393872/addicted_to_love_what_is_love_addiction_and_when_should_it_be_treated
12. Kelley AE et al., "The Neuroscience of Natural Rewards: Relevance to Addictive Drugs." *J. Neurosci.* 22, 3306–11 (2002).
13. Sussman S, "Love Addiction: Definition, Etiology, Treatment." *J. Treatment Prevention* 17, 31–45 (2010).

CHAPTER 3. DESIRE AND DOPAMINE, PLEASURE AND OPIOIDS

1. Jastreboff AM et al., "Body Mass Index, Metabolic Factors, and Striatal Activation During Stressful and Neutral-Relaxing States: An fMRI Study." *Neuropsychopharmacology* 36, 627–37 (2011).
2. Hommel JD et al., "Leptin Receptor Signaling in Midbrain Dopamine Neurons Regulates Feeding." *Neuron* 51, 801–10 (2006).
3. Farooqi IS et al., "Leptin Regulates Striatal Regions and Human Eating Behavior." *Science* 317, 1355 (2007).
4. Jastreboff AM et al., "Leptin Is Associated with Exaggerated Brain Reward and Emotion Responses to Food Images in Adolescent Obesity." *Diab. Care* 37, 3061–8 (2014).
5. Jastreboff AM et al., "Neural Correlates of Stress- and Food Cue-Induced Food Craving in Obesity: Association with Insulin Levels." *Diab. Care* 36, 394–402 (2013).
6. Rapuano KM et al., "Genetic Risk for Obesity Predicts Nucleus Accumbens Size and Responsivity to Real-World Food Cues." *Proc. Natl. Acad. Sci.* 114, 160–5 (2017).
7. Jacobs E et al., "Estrogen Shapes Dopamine-Dependent Cognitive Processes: Implications for Women's Health." *J. Neurosci.* 31, 5286–93 (2011).

8. Kenakin T, "The Mass Action Equation in Pharmacology." *Br. J. Clin. Pharmacol.* 81, 41–51 (2015).

9. Stice E et al., "Relation Between Obesity and Blunted Striatal Response to Food Is Moderated by Taq A1 Allele." *Science* 322, 449–52 (2008).

10. Girault EM et al., "Acute Peripheral but Not Central Administration of Olanzapine Induces Hyperglycemia Associated with Hepatic and Extra-Hepatic Insulin Resistance." *PLoS One* 7, e43244 (2012).

11. German CL et al., "Regulation of the Dopamine and Vesicular Monoamine Transporters: Pharmacological Targets and Implications for Disease." *Pharmacol. Rev.* 67, 1005–24 (2015).

12. Vaughan RA et al., "Mechanisms of Dopamine Transporter Regulation in Normal and Disease States." *Trends Pharmcol. Sci.* 34, 489–96 (2013).

13. Potenza MN, "How Central Is Dopamine to Pathological Gambling or Gambling Disorder?" *Front. Behav. Neurosci.* 7, 206 (2013).

14. Yip SW et al., "Health/Functioning Characteristics, Gambling Behaviors, and Gambling-Related Motivations in Adolescents Stratified by Gambling Problem Severity: Findings from a High School Survey." *Am. J. Addict.* 20, 495–508 (2011).

15. Simmons ML et al., "Endogenous Opioid Regulation of Hippocampal Function." *Int. Rev. Neurobiol.* 39, 145–96 (1996).

16. Fournier PE et al., "Effects of a 110 Kilometers Ultra-Marathon Race on Plasma Hormone Levels." *Int. J. Sports Med.* 18, 252–6 (1997).

17. Schultz W et al., "A Neural Substrate of Prediction and Reward." *Science* 275, 1593–9 (1997).

CHAPTER 4. KILLING JIMINY: STRESS, FEAR, AND CORTISOL

1. Sapolsky RM, *Why Zebras Don't Get Ulcers.* W. H. Freeman, New York (1998).

2. Kudielka BM et al., "Human Models in Acute and Chronic Stress: Assessing Determinants of Individual Hypothalamus-Pituitary-Adrenal Axis Activity and Reactivity." *Stress* 13, 1–14 (2010).

3. Rosengren A et al., "Association of Psychosocial Risk Factors with Risk of Acute Myocardial Infarction in 11,119 Cases and 13,648 Controls from 52 Countries (The INTERHEART Study): Case-Control Study." *Lancet* 364, 953–62 (2004).

4. Crump C et al., "Stress Resilience and Subsequent Risk of Type 2 Diabetes in 1.5 Million Young Men." *Diabetologia* 59, 728–33 (2016).

5. Wilson RS et al., "Chronic Psychological Distress and Risk of Alzheimer's Disease in Old Age." *Neuroepidemiology* 27, 143–63 (2006).

6. Elovainio M et al., "Socioeconomic Differences in Cardiometabolic Factors: Social Causation or Health-Related Selection? Evidence from the Whitehall II Cohort Study, 1991–2004." *Am. J. Epidemiol.* 174, 779–89 (2011).

7. Williams DR, "Race, Socioeconomic Status, and Health: The Added Effects of Racism and Discrimination." *Ann. NY Acad. Sci.* 696, 173–88 (1999).

8. Mead H et al., "Racial and Ethnic Disparities in U.S. Health Care: A Chartbook" (2008). http://www.commonwealthfund.org/usr_doc/mead_racialethnicdisparities _chartbook_1111.pdf

9. Gray JM et al., "Corticotropin-Releasing Hormone Drives Anandamide Hydrolysis in the Amygdala to Promote Anxiety." *J. Neurosci.* 35, 3879–92 (2015).

10. Meier MH et al., "Persistent Cannabis Users Show Neuropsychological Decline from Childhood to Midlife." *Proc. Natl. Acad. Sci.* 109, E2657–E2664 (2012).

11. Sapolsky RM, "Depression, Antidepressants, and the Shrinking Hippocampus." *Proc. Natl. Acad. Sci.* 98, 12320–2 (2001).

12. Finsterwald C et al., "Stress and Glucocorticoid Receptor-Dependent Mechanisms in Long-term Memory: From Adaptive Responses to Psychopathologies." *Neurobiol. Learn. Mem.* 112, 17–29 (2014).

13. Dedovic K et al., "The Brain and the Stress Axis: The Neural Correlates of Cortisol Regulation in Response to Stress." *Neuroimage* 47, 864–71 (2009).

14. Arnsten AF, "Stress Weakens Prefrontal Networks: Molecular Insults to Higher Cognition." *Nat. Neurosci.* 18, 1376–85 (2015).

15. Sellitto M et al., "The Neurobiology of Intertemporal Choice: Insight from Imaging and Lesion Studies." *Rev. Neurosci.* 22, 565–74 (2011).

16. Dallman MF et al., "Chronic Stress and Comfort Foods: Self-Medication and Abdominal Obesity." *Brain, Behavior, and Immunity* 19, 275–80 (2005).

17. Kim P et al., "Effects of Childhood Poverty and Chronic Stress on Emotion Regulatory Brain Function in Adulthood." *Proc. Natl. Acad. Sci.* 110, 18442–7 (2013).

18. Brischoux F et al., "Phasic Excitation of Dopamine Neurons in Ventral VTA by Noxious Stimuli." *Proc. Natl. Acad. Sci.* 106, 4894–9 (2009).

19. Goeders NE, "Stress and Cocaine Addiction." *J. Pharmacol. and Exp. Ther.* 301, 785–9 (2002).

20. Jupp B et al., "Social Dominance in Rats: Effects on Cocaine Self-Administration, Novelty Reactivity and Dopamine Receptor Binding and Content in the Striatum." *Psychopharmacology* 233, 579–89 (2015).

21. Nader MA et al., "Review. Positron Emission Tomography Imaging Studies of Dopamine Receptors in Primate Models of Addiction." *Philos. Trans. R. Soc. Lond. B. Biol. Sci.* 363, 3223–32 (2008).

22. Wang Y et al., "The Obesity Epidemic in the United States—Gender, Age, Socioeconomic, Racial/Ethnic, and Geographic Characteristics: A Systematic Review and Meta-Regression Analysis." *Epidemiol. Rev.* 29, 6–28 (2007).

23. Williams CT et al., "Neighborhood Socioeconomic Status, Personal Network Attributes, and Use of Heroin and Cocaine." *Am. J. Prev. Med.* 32, S203–S210 (2007).

24. Cleck JN et al., "Making a Bad Thing Worse: Adverse Effects of Stress on Drug Addiction." *J. Clin. Invest.* 118, 454–61 (2008).

25. Arnsten AF, "Prefrontal Cortical Network Connections: Key Site of Vulnerability in Stress and Schizophrenia." *Int. J. Dev. Neurosci.* 29, 215–23 (2011).

26. Cleck JN et al., "Making a Bad Thing Worse: Adverse Effects of Stress on Drug Addiction." *J. Clin. Invest.* 118, 454–61 (2008).

27. Van Huijstee AN et al., "Glutamatergic Synaptic Plasticity in the Mesocorticolimbic System in Addiction." *Front. Cell Neurosci.* 8, 466 (2015). 10.3389/Fncel.2014.00466

28. Roemmich JN et al., "Dietary Restraint and Stress-Induced Snacking in Youth." *Obes. Res.* 10, 1120–6 (2002).

29. Medic N et al., "Increased Body Mass Index Is Associated with Specific Regional Alterations in Brain Structure." *Int. J. Obes.* 40, 1177–82 (2016).

30. Dallman MF et al., "Chronic Stress and Obesity: A New View of 'Comfort Food.'" *Proceedings of the National Academy of Sciences* 100, 11696–701 (2003).
31. Pecoraro NC et al., "Chronic Stress Promotes Palatable Feeding, Which Reduces Signs of Stress: Feedforward and Feedback Effects on Chronic Stress." *Endocrinology* 145, 3754–62 (2004).
32. Gluck ME, "Stress Response and Binge Eating Disorder." *Appetite* 46, 26–30 (2006).
33. Tataranni PA et al., "Effects of Glucocorticoids on Energy Metabolism and Food Intake in Humans." *Am. J. Physiol.* 271, E317–E325 (1996).
34. Newman E et al., "Daily Hassles and Eating Behaviour: The Role of Cortisol Reactivity Status." *Psychoneuroendocrinology* 32, 125–32 (2007).
35. Adam TC et al., "Stress, Eating, and the Reward System." *Physiol. Behav.* 91, 449–58 (2007).
36. Boggiano MM et al., "Eating Tasty Foods to Cope, Enhance Reward, Socialize or Conform: What Other Psychological Characteristics Describe Each of These Motives?" *J. Health Psychol.* (2015).
37. Patel SR et al., "Association Between Reduced Sleep and Weight Gain in Women." *Am. J. Epidemiol.* 164, 947–54 (2006).
38. St-Onge MP et al., "Short Sleep Duration Increases Energy Intakes but Does Not Change Energy Expenditure in Normal-Weight Individuals." *Am. J. Clin. Nutr.* 94, 410–6 (2011).
39. Greer SM et al., "The Impact of Sleep Deprivation on Food Desire in the Human Brain." *Nat. Commun.* 4, 2259 (2013). 10.1038/Ncomms3259.
40. Waters KA et al., "Structural Equation Modeling of Sleep Apnea, Inflammation, and Metabolic Dysfunction in Children." *J. Sleep Res.* 16, 388–95 (2007).
41. Charmandari E et al., "Pediatric Stress: Hormonal Mediators and Human Development." *Horm. Res.* 59, 161–79 (2003).
42. Shonkoff JP et al., "The Lifelong Effects of Early Childhood Adversity and Toxic Stress." *Pediatrics* 129, E232–E246 (2012).
43. Greenfield EA et al., "Violence from Parents in Childhood and Obesity in Adulthood: Using Food in Response to Stress as a Mediator of Risk." *Soc. Sci. Med.* 68, 791–8 (2009).
44. Oliver G et al., "Perceived Effects of Stress on Food Choice." *Physiol. Behav.* 66, 511–5 (1999).
45 Roemmich JN et al., "Dietary Restraint and Stress-Induced Snacking in Youth." *Obes. Res.* 10, 1120–6 (2002).
46. Sinha R et al. "Stress as a Common Risk Factor for Obesity and Addiction." *Biol. Psychiatry* 73, 827–35 (2013).

CHAPTER 5. THE DESCENT INTO HADES

1. Lepousez G et al., "Adult Neurogenesis and the Future of the Rejuvenating Brain Circuits." *Neuron* 86, 387–401 (2015).
2. Gass JT et al., "Glutamatergic Substrates of Drug Addiction and Alcoholism." *Biochem. Pharmacol.* 75, 218–65 (2008).
3. De Souza L et al., "Effect of Chronic Sleep Restriction and Aging on Calcium Signaling and Apoptosis in the Hippocampus of Young and Aged Animals." *Prog. Neuropsychopharmacol. Biol. Psychiatr.* 39, 23–30 (2012).

4. Moss M, *Salt, Sugar, Fat: How the Food Giants Hooked Us*. Random House, New York (2013).

5. Heinz A et al., "Pharmacogenetic Insights to Monoaminergic Dysfunction in Alcohol Dependence." *Psychopharmacology* 174, 561–70 (2004).

6. Volkow ND et al., "Loss of Dopamine Transporters in Methamphetamine Abusers Recovers with Protracted Abstinence." *J. Neurosci.* 21, 9414–8 (2001).

7. Shmulewitz D et al., "Commonalities and Differences Across Substance Use Disorders: Phenomenological and Epidemiological Aspects." *Alcohol Clin. Exp. Res.* 39, 1878–900 (2015).

8. Glantz SA et al., *The Cigarette Papers*. University of California Press, Berkeley (1996).

9. Proctor C, "BAT Industries—Smoking Gun?" *Observer*, P13, March 1, 1998.

10. Schmidt LA, "What Are Addictive Substances and Behaviours and How Far Do They Extend?" in *Impact of Addictive Substances and Behaviours on Individual and Societal Well-Being*, Anderson P, Rehm J, Room R, eds. Oxford University Press, London (2016).

11. American Psychiatric Association, *Diagnostic and Statistical Manual of Mental Disorders: DSM-5*. American Psychiatric Association, Washington, D.C. (2013).

12. Farris SP et al., "Applying the New Genomics to Alcohol Dependence." *Alcohol* 49, 825–36 (2015).

13. Harrell PT et al., "Dopaminergic Genetic Variation Moderates the Effect of Nicotine on Cigarette Reward." *Psychopharmacology* 233, 351–60 (2015).

14. Goldman D et al., "The Genetics of Addictions: Uncovering the Genes." *Nat. Rev. Genet.* 6, 521–32 (2005).

15. Gendreau KE et al., "Detecting Associations Between Behavioral Addictions and Dopamine Agonists in the Food and Drug Administration's Adverse Event Database." *J. Behav. Addict.* 3, 21–6 (2014).

16. O'Reilly WFB, "Off Sugar, and Wanting to Tear My Eyes Out." *Newsday*, Aug. 15, 2014. http://www.newsday.com/opinion/columnists/william-f-b-o-reilly/ off-sugar-and-wanting-to-tear-my-eyes-out-william-f-b-o-reilly-1.9070131

17. King WC et al., "Prevalence of Alcohol Use Disorders Before and After Bariatric Surgery." *JAMA* 307, 2516–25 (2012).

18. Pendergrast M, *For God, Country, and Coca-Cola: The Definitive History of the Great American Soft Drink and the Company That Makes It*. Perseus, Philadelphia (2013).

19. Investor Guide, "What Are the Most Commonly Traded Commodities?" http://www .investorguide.com/article/11836/what-are-the-most-commonly-traded-commodities-igu/

CHAPTER 6. THE PURIFICATION OF ADDICTION

1. Cohen R, "Sugar Love: A Not-So-Sweet Story." *National Geographic*, Aug. 2013.

2. Schaffer Library of Drug Policy, "A Summary of Historical Events" (2015). http://www .druglibrary.org/schaffer/history/histsum.htm

3. Barras C, "Founders of Western Civilisation Were Prehistoric Dope Dealers." *New Scientist*, London (2016). https://www.newscientist.com/article/2096440-founders-of -western-civilisation-were-prehistoric-dope-dealers/?cmpid=nlc%7cnsns%7c2016-1407 -newglobal&utm_medium=nlc&utm_source=nsns

4. Lindesmith AR, "Addiction and Opiates." *Aldine Transaction* (2008).

5. Fort J, *The Pleasure Seekers: The Drug Crisis, Youth, and Society.* Grove Press, New York (1970).

6. Lindesmith AR, *The Addict and the Law.* University of Indiana Press, Bloomington (1965).

7. Drugfacts: Nationwide Trends. http://www.drugabuse.gov/publications/drugfacts /nationwide-trends

8. Statistics and Facts About the Alcoholic Beverage Industry in the U.S. http://www .statista.com/topics/1709/alcoholic-beverages/

9. Keyes KM et al., "National Multi-Cohort Time Trends in Adolescent Risk Preference and the Relation with Substance Use and Problem Behavior from 1976 to 2011." *Drug Alcohol Depend.* 155, 267–74 (2015).

10. Jardin B et al., "Characteristics of College Students with Attention-Deficit Hyperactivity Disorder Symptoms Who Misuse Their Medications." *J. Am. Coll. Health* 59, 373–7 (2011).

11. Pharmaceutical Industry Gets High on Fat Profits. http://www.bbc.com/news /business-28212223

12. Stevens B et al., "Sucrose for Analgesia in Newborn Infants Undergoing Painful Procedures." *Cochrane Database Syst. Rev.* 7, CD001069 (2016).

13. Staff of the Select Committee on Nutrition and Human Needs, U.S. Senate, *Dietary Goals for the United States.* U.S. Government Printing Office, Washington, D.C. (1977). https://naldc.nal.usda.gov/naldc/download.xhtml?id=1759572&content=pdf

14. Wallace DL et al., "The Influence of DeltaFosB in the Nucleus Accumbens on Natural Reward-Related Behavior." *J. Neurosci.* 28, 10272–7 (2008).

15. Colantuoni C et al., "Evidence That Intermittent Excessive Sugar Intake Causes Endogenous Opioid Dependence." *Obes. Res.* 10, 478–88 (2002).

16. Stice E et al., "Relative Ability of Fat and Sugar Tastes to Activate Reward, Gustatory, and Somatosensory Regions." *Am. J. Clin. Nutr.* 98, 1377–84 (2013).

17. Lustig RH, "Fructose: It's Alcohol Without the 'Buzz.'" *Adv. Nutr.* 4, 226–35 (2013).

18. Wölnerhanssen BK et al., "Dissociable Behavioral, Physiological and Neural Effects of Acute Glucose and Fructose Ingestion: A Pilot Study." *PLoS One* 10, e0130280 (2015).

19. Alcohol, Tobacco, and Other Drugs. https://www.samhsa.gov/atod

20. Lustig RH, "Fructose: It's Alcohol Without the 'Buzz.'" *Adv. Nutr.* 4, 226–35 (2013).

21. Lustig RH et al., "The Toxic Truth About Sugar." *Nature* 487, 27–9 (2012).

22. Lindqvist A et al., "Effects of Sucrose, Glucose and Fructose on Peripheral and Central Appetite Signals." *Regul. Pept.* 150, 26–32 (2008).

23. Pelchat ML et al., "Images of Desire: Food-Craving Activation During fMRI." *Neuroimage* 23, 1486–93 (2004).

24. Lenoir M et al., "Intense Sweetness Surpasses Cocaine Reward." *PLoS One* 2, e698 (2007).

25. Avena NM et al., "Evidence for Sugar Addiction: Behavioral and Neurochemical Effects of Intermittent, Excessive Sugar Intake." *Neurosci. Biobehav. Rev.* 32, 20–39 (2008).

26. Hebebrand J et al., "'Eating Addiction,' Rather Than 'Food Addiction,' Better Captures Addictive-Like Eating Behavior." *Neurosci. Biobehav. Rev.* 47, 295–300 (2014).

27. Neurofast, "Neurofast Consensus Opinion on Food Addiction" (2014). http://www .neurofast.eu/consensus

28. Pesis E, "The Role of the Anaerobic Metabolites, Acetaldehyde and Ethanol, in Fruit Ripening, Enhancement of Fruit Quality and Fruit Deterioration." *Postharvest Biol. Technol.* 37, 1–19 (2005).
29. Whelton PK et al., "Sodium, Blood Pressure, and Cardiovascular Disease: Further Evidence Supporting the American Heart Association Sodium Reduction Recommendations." *Circulation* 126, 2880–9 (2012).

CHAPTER 7. CONTENTMENT AND SEROTONIN

1. U.S. Burden of Disease Collaborators, "The State of U.S. Health, 1990–2010: Burden of Diseases, Injuries, and Risk Factors." *JAMA* 310, 591–608 (2013).
2. Szalavitz M, "What Does a 400% Increase in Antidepressant Use Really Mean?" *Time*, Oct. 20, 2011. http://healthland.time.com/2011/10/20/what-does-a-400-increase-in-antidepressant-prescribing-really-mean/
3. Gu Q et al., "Prescription Drug Use Continues to Increase: U.S. Prescription Drug Data for 2007–2008." National Center for Health Statistics, Centers for Disease Control (2010). http://www.cdc.gov/nchs/data/databriefs/db42.pdf
4. Stone K, "The Most Prescribed Medications by Drug Class." *Balance*, Oct. 13, 2016. http://pharma.about.com/od/sales_and_marketing/a/the-most-prescribed-medications-by-drug-class.htm
5. Gu Q et al., "Prescription Drug Use Continues to Increase: U.S. Prescription Drug Data for 2007–2008." National Center for Health Statistics, Centers for Disease Control (2010). http://www.cdc.gov/nchs/data/databriefs/db42.pdf
6. Case BG et al., "Trends in the Inpatient Mental Health Treatment of Children and Adolescents in US Community Hospitals Between 1990 and 2000." *Arch. Gen. Psychiatry* 64, 89–96 (2007).
7. Berger M et al., "The Expanded Biology of Serotonin." *Ann. Rev. Med.* 60, 355–66 (2009).
8. Mann JJ et al., "The Neurobiology and Genetics of Suicide and Attempted Suicide: A Focus on the Serotonergic System." *Neuropsychopharmacol.* 24, 467–77 (2001).
9. Heisler LK et al., "Elevated Anxiety and Antidepressant-Like Responses in Serotonin 5-HT$_{1A}$ Receptor Mutant Mice." *Proc. Natl. Acad. Sci.* 95, 15049–54 (1998).
10. Dass JF et al., "Computational Exploration of Polymorphisms in 5-Hydoxytryptamine 5-HT$_{1A}$ and 5-HT$_{2A}$ Receptors Associated with Psychiatric Disease." *Gene* 502, 16–26 (2012).
11. Kishi T et al., "Serotonin 1A Receptor Gene, Schizophrenia and Bipolar Disorder: an Association Study and Meta-Analysis." *Psychiatr. Res.* 185, 20–6 (2011).
12. Köhler S et al., "The Serotonergic System in the Neurobiology of Depression: Relevance for Novel Antidepressants." *J. Psychopharmacol.* 30, 13–22 (2015).
13. Chilmonczyk Z et al., "Functional Selectivity and Antidepressant Activity of Serotonin 1A Receptor Ligands." *Int. J. Mol. Sci.* 16, 18474–506 (2015).
14. Blier P et al., "Serotonin and Beyond: Therapeutics for Major Depression." *Philos. Trans. R. Soc. Lond. B Biol. Sci.* 368, 20120536 (2013).
15. Clauss JA et al., "The Nature of Individual Differences in Inhibited Temperament and Risk for Psychiatric Disease: A Review and Meta-Analysis." *Prog. Neurobiol.* 127–28, 23–45 (2015).

16. Murphy DL et al., "Targeting the Murine Serotonin Transporter: Insights into Human Biology." *Nat. Rev. Neurosci.* 9, 85-96 (2008).

17. Karg K et al. "The Serotonin Transporter Promoter Variant (5HTTLPR), Stress, and Depression Meta-Analysis Revisited: Evidence of Genetic Moderation." *Arch. Gen. Psychiatry* 68, 444–54 (2011).

18. Barnes DM et al., "Racial Differences in Depression in the United States: How Do Subgroup Analyses Inform a Paradox?" *Soc. Psychiatr. Psych. Epidemiol.* 48, 1941–9 (2013).

19. Himle JA et al., "Anxiety Disorders Among African Americans, Blacks of Caribbean Descent, and Non-Hispanic Whites in the United States." *J. Anxiety Dis.* 23, 578–90 (2009).

20. Williams MT et al., "The Role of Ethnic Identity in Symptoms of Anxiety and Depression in African Americans." *Psychiatr. Res.* 199, 31–6 (2012).

21. Odgerel Z et al., "Genotyping Serotonin Transporter Polymorphisms 5-HTTLPR and Rs25531 in European- and African-American Subjects from the National Institute of Mental Health's Collaborative Center for Genomic Studies." *Trans. Psychiatry* 3, E307 (2013).

22. Coccaro EF et al., "Serotonin and Impulsive Aggression." *CNS Spectrum* 20, 295–302 (2015).

23. Iqbal MM et al., "Overview of Serotonin Syndrome." *Ann. Clin. Psychiatry* 24, 310–8 (2012).

24. Young SN, "How to Increase Serotonin in the Brain Without Drugs." *Rev. Psychiatr. Neurosci.* 32, 394–9 (2007).

25. Odgerel Z et al., "Genotyping Serotonin Transporter Polymorphisms 5-HTTLPR and Rs25531 in European- and African-American Subjects from the National Institute of Mental Health's Collaborative Center for Genomic Studies." *Trans. Psychiatry* 3, E307 (2013).

CHAPTER 8. PICKING THE LOCK TO NIRVANA

1. Hofmann A, "How LSD Originated." *J. Psychedelic Drugs* 11, 53–60 (1979).

2. Twarog BM et al., "Serotonin Content of Some Mammalian Tissues and Urine and a Method for Its Determination." *Am. J. Physiol.* 175, 157–61 (1953).

3. Grof S, *Realms of the Human Unconscious: Observations from LSD Research.* Viking, New York (1975).

4. Szalavitz M, "The Legacy of the CIA's Secret LSD Experiments on America," *Time*, March 23, 2012. http://healthland.time.com/2012/03/23/the-legacy-of-the-cias-secret-lsd-experiments-on-america/

5. Stone AL et al., "Evidence for a Hallucinogen Dependence Syndrome Developing Soon After Onset of Hallucinogen Use During Adolescence." *Int. J. Methods Psychiatr. Res.* 15, 116–30 (2006).

6. Nichols DE, "Psychedelics." *Pharmacol. Rev.* 68, 264–355 (2016).

7. Krebs TS et al., "Psychedelics and Mental Health: A Population Study." *PLoS One* 8, e63972 (2013).

8. Griffiths RR et al., "Psilocybin Occasioned Mystical-Type Experiences: Immediate and Persisting Dose-Related Effects." *Psychopharmacology* 218, 649–65 (2011).

9. Giedd JN, "Structural Magnetic Resonance Imaging of the Adolescent Brain." *Ann. NY Acad. Sci.* 1021, 77–85 (2004).

10. Pollan M, "The Trip Treatment." *New Yorker*, Feb. 9, 2015. http://www.newyorker.com /magazine/2015/02/09/trip-treatment

11. Grob CS et al., "Pilot Study of Psilocybin Treatment for Anxiety in Patients with Advanced-Stage Cancer." *Arch. Gen. Psychiatry* 68, 71–8 (2011).

12. Nichols DE, "Psychedelics." *Pharmacol. Rev.* 68, 264–355 (2016).

13. Gasser P et al., "Safety and Efficacy of Lysergic Acid Diethylamide-Assisted Psychotherapy for Anxiety Associated with Life-Threatening Diseases." *J. Nerv. Ment. Dis.* 202, 513–20 (2014).

14. Griffiths RR et al., "Psilocybin Occasioned Mystical-Type Experiences: Immediate and Persisting Dose-Related Effects." *Psychopharmacology* 218, 649–65 (2011).

15. Krebs TS et al., "Lysergic Acid Diethylamide (LSD) for Alcoholism: Meta-Analysis of Randomized Controlled Trials." *J. Psychopharmacol.* 26, 994–1002 (2012).

16. Stahl SM, "Mechanism of Action of Serotonin Selective Reuptake Inhibitors. Serotonin Receptors and Pathways Mediate Therapeutic Effects and Side Effects." *J. Affect. Disord.* 51, 215–35 (1998).

17. Chilmonczyk Z et al., "Functional Selectivity and Antidepressant Activity of Serotonin 1A Receptor Ligands Int." *J. Mol. Sci.* 16, 18474–506 (2015).

18. Heisler LK et al. "Elevated Anxiety and Antidepressant-Like Responses in Serotonin 5-HT$_{1A}$ Receptor Mutant Mice." *Proc. Natl. Acad. Sci.* 95, 15049–54 (1998).

19. Kishi T et al., "The Serotonin 1A Receptor Gene Confer Susceptibility to Mood Disorders: Results from an Extended Meta-Analysis of Patients with Major Depression and Bipolar Disorder." *Eur. Arch. Psychiatry Clin. Neurosci.* 263, 105–18 (2013).

20. Carhart-Harris RL et al., "Neural Correlates of the LSD Experience Revealed by Multimodal Neuroimaging." *Proc. Natl. Acad. Sci.* 113, 4853–8 (2016).

21. Vollenweider FX et al., "Positron Emission Tomography and Fluorodeoxyglucose Studies of Metabolic Hyperfrontality and Psychopathology in the Psilocybin Model of Psychosis." *Neuropsychopharmacology* 16, 357-72 (1997).

22. Nichols DE, "Psychedelics." *Pharmacol. Rev.* 68, 264–355 (2016).

23. Halberstadt AL et al., "Differential Contributions of Serotonin Receptors to the Behavioral Effects of Indoleamine Hallucinogens in Mice." *J. Psychopharmacol.* 25, 1548–61 (2011).

24. Majic T et al., "Peak Experiences and the Afterglow Phenomenon: When and How Do Therapeutic Effects of Hallucinogens Depend on Psychedelic Experiences?" *J. Psychopharmacol.* 29, 241–53 (2015).

25. Blough BE et al., "Interaction of Psychoactive Tryptamines with Biogenic Amine Transporters and Serotonin Receptor Subtypes." *Psychopharmacology* 231, 4135–44 (2014).

26. Fantegrossi WE et al., "The Behavioral Pharmacology of Hallucinogens." *Biochem. Pharmacol.* 75, 17–33 (2008).

27. Monte AP et al., "Dihydrobenzofuran Analogues of Hallucinogens. 4. Mescaline Derivatives." *J. Med. Chem.* 40, 2997–3008 (1997).

28. Halberstadt AL et al., "Multiple Receptors Contribute to the Behavioral Effects of Indoleamine Hallucinogens." *Neuropharmacology* 61, 364–81 (2011).

29. Carhart-Harris RL et al., "LSD Enhances Suggestibility in Healthy Volunteers." *Psychopharmacology* 232, 785–94 (2015).

30. Maclean KA et al., "Mystical Experiences Occasioned by the Hallucinogen Psilocybin Lead to Increases in the Personality Domain of Openness." *J. Psychopharmacol.* 25, 1453–61 (2011).

31. Bogenschutz MP et al., "Classic Hallucinogens in the Treatment of Addictions." *Prog. Neuropsychopharmacol. Biol. Psychiatry* 64, 250–8 (2016).

32. Cholden LS et al., "Clinical Reactions and Tolerance to LSD in Chronic Schizophrenia." *J. Nerv. Ment. Dis.* 112, 211–21 (1955).

33. Buckholtz NS et al. "Lysergic Acid Diethylamide (LSD) Administration Selectively Downregulates Serotonin2 Receptors in Rat Brain." *Neuropsychopharmacology* 3, 137–48 (1990).

CHAPTER 9. WHAT YOU EAT IN PRIVATE YOU WEAR IN PUBLIC

1. Meerlo P et al., "Restricted and Disrupted Sleep: Effects on Autonomic Function, Neuroendocrine Stress Systems and Stress Responsivity." *Sleep Med. Rev.* 12, 197–210 (2008).

2. Silber BY et al., "Effect of Tryptophan Loading on Human Cognition, Mood, and Sleep." *Neurosci. Behav. Rev.* 34, 387–407 (2010).

3. Bhatti T et al., "Effects of a Tryptophan-Free Amino Acid Drink Challenge on Normal Human Sleep Electroencephalogram and Mood." *Biol. Psychiatry* 43, 52–9 (1998).

4. Franckle RL et al., "Insufficient Sleep Among Elementary and Middle School Students Is Linked with Elevated Soda Consumption and Other Unhealthy Dietary Behaviors." *Prev. Med.* 74, 36–41 (2015).

5. Pongpaew P et al., "The Nutritional Availability of Tryptophan in Foods." *Eur. Rev. Nutr. Diet.* 10, 297–308 (1968).

6. Deshazo RD et al., "The Autopsy of Chicken Nuggets Reads 'Chicken Little.'" *Am. J. Med.* 126, 1018–9 (2016).

7. Schmidt JA et al., "Plasma Concentrations and Intakes of Amino Acids in Male Meat-Eaters, Fish-Eaters, Vegetarians and Vegans: A Cross-Sectional Analysis in the EPIC-Oxford Cohort." *Eur. J. Clin. Nutr.* 102, 1518–26 (2015).

8. Badawy AA, "Plasma Free Tryptophan Revisited: What You Need to Know and Do Before Measuring It." *J. Psychopharmacol.* 24, 809–15 (2010).

9. Li F et al., "Fish Consumption and Risk of Depression: a Meta-Analysis." *J. Epidemiol. Community Health* 70, 299–304 (2016).

10. Mohajeri MH et al., "Chronic Treatment with a Tryptophan-Rich Protein Hydrolysate Improves Emotional Processing, Mental Energy Levels and Reaction Time in Middle-Aged Women." *Br. J. Nutr.* 9, 1–16 (2015).

11. Jangid P et al., "Comparative Study of Efficacy of L-5-Hydroxytryptophan and Fluoxetine in Patients Presenting with First Depressive Episode." *Asian J. Psychiatry* 6, 29–34 (2013).

12. Aan Het Rot M et al., "Social Behaviour and Mood in Everyday Life: Effects of Tryptophan in Quarrelsome Individuals." *J. Psychiatry Neurosci.* 31, 253–62 (2006).

13. Parker G et al., "Mood Effects of the Amino Acids Tryptophan and Tyrosine: 'Food for Thought' III." *Acta Psychiatr. Scand.* 124, 417–26 (2011).

14. Li X et al., "Composition of Amino Acids in Feed Ingredients for Animal Diets." *Amino Acids* 40, 1159–68 (2011).

15. Uhe AM et al., "A Comparison of the Effects of Beef, Chicken and Fish Protein on Satiety and Amino Acid Profiles in Lean Male Subjects." *J. Nutr.* 122, 467–72 (1992).

16. Layman DK et al., "Potential Importance of Leucine in Treatment of Obesity and the Metabolic Syndrome." *J. Nutr.* 136 (1Supp), 319S-23S (2006).

17. Tai ES et al., "Insulin Resistance Is Associated with a Metabolic Profile of Altered Protein Metabolism in Chinese and Asian-Indian Men." *Diabetologia* 53, 757–87 (2010).

18. Batch BC et al., "Branch Chain Amino Acids: Biomarkers of Health and Disease." *Curr. Opin. Clin. Nutr. Metab. Care* 17, 86–9 (2014).

19. Muldoon MF et al., "The Metabolic Syndrome Is Associated with Reduced Central Serotonergic Responsivity in Healthy Community Volunteers." *J. Clin. Endocrinol. Metab.* 91, 718–21 (2006).

20. Thomas EL et al., "The Missing Risk: MRI and MRS Phenotyping of Abdominal Adiposity and Ectopic Fat." *Obesity* 20, 76–87 (2012).

21. Fabbrini E et al., "Intrahepatic Fat, Not Visceral Fat, Is Linked with Metabolic Complications of Obesity." *Proc. Natl. Acad. Sci.* 106, 15430–5 (2009).

22. Liaw FY et al., "Exploring the Link Between the Components of Metabolic Syndrome and the Risk of Depression." *Biomed. Res. Int.* doi: 10.1155/2015/586251 (2015).

23. Horvath JD et al., "Food Consumption in Patients Referred for Bariatric Surgery with and Without Binge Eating Disorder." *Eat. Behav.* 19, 173–6 (2015).

24. Shai I et al., "Weight Loss with a Low-Carbohydrate, Mediterranean, or Low-Fat Diet." *N. Engl. J. Med.* 359, 229–41 (2008).

25. García-Toro M et al., "Obesity, Metabolic Syndrome and Mediterranean Diet: Impact on Depression Outcome." *J. Affect. Disord.* 194, 105–8 (2016).

26. Chung CC et al., "Inflammation-Associated Declines in Cerebral Vasoreactivity and Cognition in Type 2 Diabetes." *Neurology* 85, 450–8 (2015).

27. De La Monte S, "Brain Insulin Resistance and Deficiency as Therapeutic Targets in Alzheimer's Disease." *Curr. Alzheimer's Res.* 9, 35–66 (2012).

28. Bitel CL et al., "Amyloid-ß and Tau Pathology of Alzheimer's Disease Induced by Diabetes in a Rabbit Animal Model." *J. Alzheimer's Dis.* 32, 291–305 (2012).

29. Lustig RH, "Childhood Obesity: Behavioral Aberration or Biochemical Drive? Reinterpreting the First Law of Thermodynamics." *Nature Clin. Pract. Endo. Metab.* 2, 447–58 (2006).

30. Klöckener T et al., "High-Fat Feeding Promotes Obesity Via Insulin Receptor/PI3K-Dependent Inhibition of SF-1 VMH Neurons." *Nat. Neurosci.* 14, 911–8 (2011).

31. Talbot K et al., "Demonstrated Brain Insulin Resistance in Alzheimer's Disease Patients Is Associated with IGF-1 Resistance, IRS-1 Dysregulation, and Cognitive Decline." *J. Clin. Invest.* 122, 1316–38 (2012).

32. Cholerton B et al. "Insulin, Cognition, and Dementia." *Eur. J. Pharmacol.* 719, 170–9 (2013).

33. Yau PL et al., "Obesity and Metabolic Syndrome and Functional and Structural Brain Impairments in Adolescence." *Pediatrics* 130, E856–E864 (2012).

34. Lakhan SE et al., "The Emerging Role of Dietary Fructose in Obesity and Cognitive Decline." *Nutr. J.* 12, 114 (2013).

35. Orr ME et al., "Mammalian Target of Rapamycin Hyperactivity Mediates the Detrimental Effects of a High Sucrose Diet on Alzheimer's Disease Pathology." *Neurobiol. Aging* 35, 1233–42 (2014).

36. Cisternas P et al., "Fructose Consumption Reduces Hippocampal Synaptic Plasticity Underlying Cognitive Performance." *Biochim. Biophys. Acta* 1852, 2379–90 (2015).

37. Meng Q et al., "Systems Nutrigenomics Reveals Brain Gene Networks Linking Metabolic and Brain Disorders." *E-Biomedicine* (2016). http://dx.doi.org/10.1016/j.ebiom.2016.04.008

38. Lakhan SE et al., "The Emerging Role of Dietary Fructose in Obesity and Cognitive Decline." *Nutr. J.* 12, 114 (2013).

39. Seneff S et al., "Nutrition and Alzheimer's Disease: The Detrimental Role of a High Carbohydrate Diet." *Eur. J. Intern. Med.* 22, 134–40 (2011).

40. Westover AN et al., "A Cross-National Relationship Between Sugar Consumption and Depression?" *Dep. Anxiety* 16, 118–20 (2002).

41. Peet M, "International Variations in the Outcome of Schizophrenia and the Prevalence of Depression in Relation to National Dietary Practices: an Ecological Analysis." *Br. J. Psychiatry* 184, 404–8 (2004).

42. Gangwisch JE et al., "High Glycemic Index Diet as a Risk Factor for Depression: Analyses from the Women's Health Initiative." *Am. J. Clin. Nutr.* 102, 454–63 (2015).

43. Alcock J et al., "Is Eating Behavior Manipulated by the Gastrointestinal Microbiota? Evolutionary Pressures and Potential Mechanisms." *Bioessays* 36, 940–9 (2014).

44. Foster JA et al., "Gut Microbiota and Brain Function: An Evolving Field in Neuroscience." *Int. J. Neuropsychopharmacol.* 19(5), pii: pyv114, doi: 10.1093/ijnp/pyv114 (2016).

45. Meadow JF et al., "Humans Differ in Their Personal Microbial Cloud." *PeerJ.* 3, E1258, doi: 10.7717/peerj.1258 (2015).

46. David LA et al., "Diet Rapidly and Reproducibly Alters the Human Gut Microbiome." *Nature* 505, 559–63 (2014).

47. Hoffmann C et al., "Archaea and Fungi of the Human Gut Microbiome: Correlations with Diet and Bacterial Residents." *PLoS One* 8, e66109 (2013).

48. Vijay-Kumar M et al., "Metabolic Syndrome and Altered Gut Microbiota in Mice Lacking Toll-Like Receptor 5." *Science* 328, 228–31 (2010).

49. Alang N et al., "Weight Gain After Fecal Microbiota Transplantation." *Open Forum Infect. Dis.* 2, doi:10.1093/ofid/ofv004 (2015).

50. Angelakis E et al., "Related Actions of Probiotics and Antibiotics on Gut Microbiota and Weight Modification." *Lancet Infect. Dis.* 13, 889–99 (2013).

51. Martinez RC et al., "Scientific Evidence for Health Effects Attributed to the Consumption of Probiotics and Prebiotics: An Update for Current Perspectives and Future Challenges." *Br. J. Nutr.* 114, 1993–2015 (2015).

52. Lyte M, "Microbial Endocrinology: The Microbiota-Gut-Brain Axis," in *Health and Disease*, vol. 817, Lyte M and Cryan JF, eds. Spring, New York (2014), 3–24.

53. Sommer F et al., "The Gut Microbiota—Masters of Host Development and Physiology." *Nat. Rev. Microbiol.* 11, 227–38 (2013).

54. Cladis DP et al., "Fatty Acid Profiles of Commercially Available Finfish Fillets in the United States." *Lipids* 49, 1005–18 (2014).

55. Agrawal R et al., "Metabolic Syndrome in the Brain: Deficiency in Omega-3-Fatty Acid Exacerbates Dysfunctions in Insulin Receptor Signaling and Cognition." *J. Physiol.* 590 (Pt.10), 2485–99 (2012).

56. Patrick RP et al., "Vitamin D and the Omega-3 Fatty Acids Control Serotonin Synthesis and Action, Part 2: Relevance for ADHD, Bipolar Disorder, Schizophrenia, and Impulsive Behavior." *FASEB J.* 29, 2207–22 (2015).

57. Suarez EC et al., "The Relation of Severity of Depressive Symptoms to Monocyte-Associated Proinflammatory Cytokines and Chemokines in Apparently Healthy Men." *Psychosom. Med.* 65, 362–8 (2003).

58. Vedin I et al., "Reduced Prostaglandin F2 Alpha Release from Blood Mononuclear Leukocytes After Oral Supplementation of Omega3 Fatty Acids: The OmegAD Study." *J. Lipid Res.* 51, 1179–85 (2010).

59. Maes M et al., "Fatty Acids, Cytokines, and Major Depression." *Biol. Psychiatry* 43, 313–4 (1998).

60. Wood JT et al., "Dietary Docosahexaenoic Acid Supplementation Alters Select Physiological Endocannabinoid-System Metabolites in Brain and Plasma." *J. Lipid Res.* 51, 1416–23 (2010).

61. Drug Facts: Nationwide Trends. http://www.drugabuse.gov/publications/drugfacts/nationwide-trends

62. Lafourcade M et al., "Nutritional Omega-3 Deficiency Abolishes Endocannabinoid-Mediated Neuronal Functions." *Nat. Neurosci.* 14, 345–50 (2011).

63. Drug Facts: Nationwide Trends. http://www.drugabuse.gov/publications/drugfacts/nationwide-trends

64. Jazayeri S et al., "Comparison of Therapeutic Effects of Omega-3 Fatty Acid Eicosapentaenoic Acid and Fluoxetine, Separately and in Combination, in Major Depressive Disorder." *Aust. NZ J. Psychiatry* 42, 192–8 (2008).

65. Hallahan B et al., "Omega-3 Fatty Acid Supplementation in Patients with Recurrent Self-Harm. Single-Centre Double-Blind Randomised Controlled Trial." *Br. J. Psychiatry* 190, 118–22 (2007).

66. Raine A et al., "Nutritional Supplementation to Reduce Child Aggression: a Randomized, Stratified, Single-Blind, Factorial Trial." *J. Child Psychol. Psychiatr.* 57, 1038–46 (2016).

67. Nemets H et al., "Omega-3 Treatment of Childhood Depression: A Controlled, Double-Blind Pilot Study." *Am. J. Psychiatry* 163, 1098–1100 (2006).

68. Sublette ME et al., "Meta-Analysis of the Effects of Eicosapentaenoic Acid (EPA) in Clinical Trials in Depression." *J. Clin. Psychiatry* 72, 1577–84 (2011).

69. Freeman MP et al., "Omega-3 Fatty Acids: Evidence Basis for Treatment and Future Research in Psychiatry." *J. Clin. Psychiatry* 67, 1954–67 (2006).

70. Bhatia HS et al., "Omega-3 Fatty Acid Deficiency During Brain Maturation Reduces Neuronal and Behavioral Plasticity in Adulthood." *PLoS One* 6, e28451 (2011).

71. Hibbeln JR et al., "Maternal Seafood Consumption in Pregnancy and Neurodevelopmental Outcomes in Childhood (ALSPAC Study): An Observational Cohort Study." *Lancet* 369, 579–85 (2007).

72. Makrides M et al., "Effect of DHA Supplementation During Pregnancy on Maternal Depression and Neurodevelopment of Young Children: A Randomized Controlled Trial." *JAMA* 304, 1675–83 (2010).

73. Pribis P, "Effects of Walnut Consumption on Mood in Young Adults: A Randomized Controlled Trial." *Nutrients* 8 (11), E668 (2016).

CHAPTER 10. SELF-INFLICTED MISERY:
THE DOPAMINE-CORTISOL-SEROTONIN CONNECTION

1. Kishi T et al., "The Serotonin 1A Receptor Gene Confer Susceptibility to Mood Disorders: Results from an Extended Meta-Analysis of Patients with Major Depression and Bipolar Disorder." *Eur. Arch. Psychiatry Clin. Neurosci.* 263, 105–18 (2013).

2. Bhagwagar Z et al., "Persistent Reduction in Brain Serotonin1a Receptor Binding in Recovered Depressed Men Measured by Positron Emission Tomography with [11C]-WAY-100635." *Mol. Psychiatry* 9, 386–92 (2004).

3. Savitz J et al., "5-HT(1A) Receptor Function in Major Depressive Disorder." *Prog. Neurobiol.* 88, 17–31 (2009).

4. Stahl SM, "Mechanism of Action of Serotonin Selective Reuptake Inhibitors. Serotonin Receptors and Pathways Mediate Therapeutic Effects and Side Effects." *J. Affect. Disord.* 51, 215–35 (1998).

5. Blier P et al., "Serotonin and Beyond: Therapeutics for Major Depression." *Philos. Trans. R. Soc. Lond. B. Biol. Sci.* 368, 20120536 (2013).

6. Blier P et al., "Modification of 5-HT Neuron Properties by Sustained Administration of the 5-HT1A Agonist Gepirone: Electrophysiological Studies in the Rat Brain." *Synapse* 1, 470–80 (1987).

7. Celada P et al., "Serotonin 5-HT1A Receptors as Targets for Agents to Treat Psychiatric Disorders: Rationale and Current Status of Research." *CNS Drugs* 27, 703–16 (2013).

8. Di Giovanni G et al., "Serotonin-Dopamine Interaction: An Overview." *Prog. Brain Res.* 172, 3–6 (2008).

9. De Simoni MG et al., "Modulation of Striatal Dopamine Metabolism by the Activity of Dorsal Raphe Serotonergic Afferences." *Brain Res.* 411, 81–8 (1987).

10. Müller CP et al., "The Role of Serotonin in Drug Use and Addiction." *Behav. Brain Res.* 277, 146–92 (2015).

11. Harvey-Lewis C et al., "The 5-HT(2C) Receptor Agonist Lorcaserin Reduces Cocaine Self-Administration, Reinstatement of Cocaine-Seeking and Cocaine Induced Locomotor Activity." *Neuropharmacology* 101, 237–45 (2016).

12. Nakamura K, "The Role of the Dorsal Raphé Nucleus in Reward-Seeking Behavior." *Front. Integr. Neurosci.* 7, 60 doi: 210.3389/Fnint.2013.00060 (2013).

13. Marcinkiewcz CA, "Serotonergic Systems in the Pathophysiology of Ethanol Dependence: Relevance to Clinical Alcoholism." *ACS Chem. Neurosci.* 15, 1026–39 (2015).

14. Rogers RD, "The Roles of Dopamine and Serotonin in Decision Making: Evidence from Pharmacological Experiments in Humans." *Neuropsychopharmacology* 36, 114–32 (2011).

15. Perret G et al., "Downregulation of 5-HT1A Receptors in Rat Hypothalamus and Dentate Gyrus After 'Binge' Pattern Cocaine Administration." *Synapse* 30, 166–71 (1998).

16. Rickli A et al., "Receptor Interaction Profiles of Novel Psychoactive Tryptamines Compared with Classic Hallucinogens." *Eur. Neuropsychopharmacol.* 26, 1327–37 (2016).

17. Lyvers M et al., "Illicit Use of LSD or Psilocybin, but Not MDMA or Nonpsychedelic Drugs, Is Associated with Mystical Experiences in a Dose-Dependent Manner." *J. Psychoactive Drugs* 44, 410–7 (2012).
18. Li IH et al., "Involvement of Autophagy Upregulation in 3,4-Methylenedioxymethamphetamine ('Ecstasy')–Induced Serotonergic Neurotoxicity." *Neurotoxicology* 52, 114–26 (2016).
19. Downey LA et al., "Altered Energy Production, Lowered Antioxidant Potential, and Inflammatory Processes Mediate CNS Damage Associated with Abuse of the Psychostimulants MDMA and Methamphetamine." *Eur. J. Pharmacol.* 727, 125-9 (2014).
20. Marshall JF et al. "Methamphetamine-Induced Neural and Cognitive Changes in Rodents." *Addiction* 102 Suppl. 1:61-9, 61-9 (2007).
21. Danforth AL et al., "MDMA-Assisted Therapy: A New Treatment Model for Social Anxiety in Autistic Adults." *Prog. Neuropsychopharmacol. Biol. Psychiatry* 64, 237–49 (2016).
22. Nikolaus S et al., "Cortical GABA, Striatal Dopamine and Midbrain Serotonin as the Key Players in Compulsive and Anxiety Disorders—Results from in *Vivo* Imaging Studies." *Rev. Neurosci.* 21, 119–39 (2010).
23. Cools R et al., "Inverted-U-Shaped Dopamine Actions on Human Working Memory and Cognitive Control." *Biol. Psychiatry* 69, E113-E125 (2011).
24. Bethea CL et al., "Serotonin-Related Gene Expression in Female Monkeys with Individual Sensitivity to Stress." *Neuroscience* 132, 151–66 (2005).
25. Medeiros LR et al., "Cortisol-Mediated Downregulation of the Serotonin 1A Receptor Subtype in the Gulf Toadfish, *Opsanus Beta*." *Comp. Biochem. Physiol. A Mol. Integr. Physiol.* 164, 612–21 (2013).
26. Zhong P et al., "Transcriptional Regulation of Hippocampal 5-HT1a Receptors by Corticosteroid Hormones." *Brain Res. Mol. Brain Res.* 29, 23–34 (1995).
27. Flügge G et al., "5HT1A-Receptors and Behaviour Under Chronic Stress: Selective Counteraction by Testosterone." *Eur. J. Neurosci.* 10, 2685–93 (1998).
28. Nugent AC et al., "Reduced Post-Synaptic Serotonin Type 1A Receptor Binding in Bipolar Depression." *Eur. Neuropsychopharmacol.* 23, 822–9 (2013).
29. Pompili M et al., "The Hypothalamic-Pituitary-Adrenal Axis and Serotonin Abnormalities: A Selective Overview for the Implications of Suicide Prevention." *Eur. Arch. Psychiatry Clin. Neurosci.* 260, 583–600 (2010).
30. Roy M et al., "Molecular and Genetic Basis of Depression." *J. Genet.* 93, 879–92 (2014).
31. Slopen N et al., "Interventions to Improve Cortisol Regulation in Children: A Systematic Review." *Pediatrics* 133, 312–26 (2014).
32. Schalinski I et al., "The Cortisol Paradox of Trauma-Related Disorders: Lower Phasic Responses but Higher Tonic Levels of Cortisol Are Associated with Sexual Abuse in Childhood." *PLoS One* 10, e0136921 (2015).
33. Caspi A et al., "Influence of Life Stress on Depression: Moderation by a Polymorphism in the 5-HTT Gene." *Science* 301, 386–9 (2003).
34. "Unhealthy Sleep-Related Behaviors—12 States." 2009. http://www.cdc.gov/mmwr/pdf/wk/mm6008.pdf
35. Kamphuis J et al., "Poor Sleep as a Potential Causal Factor in Aggression and Violence." *Sleep Med.* 13, 327–34 (2012).

36. Singareddy RK et al., "Sleep and Suicide in Psychiatric Patients." *Ann. Clin. Psychiatry* 13, 93–101 (2001).

37. Breslau N et al., "Sleep Disturbance and Psychiatric Disorders: A Longitudinal Epidemiological Study of Young Adults." *Biol. Psychiatry* 39, 411–8 (1996).

38. Evrard A et al., "Glucocorticoid Receptor-Dependent Desensitization of 5-HT1A Autoreceptors by Sleep Deprivation: Studies in GR-I Transgenic Mice." *Sleep* 29, 31-6 (2006).

39. Novati A et al., "Chronically Restricted Sleep Leads to Depression-Like Changes in Neurotransmitter Receptor Sensitivity and Neuroendocrine Stress Reactivity in Rats." *Sleep* 31, 1579–85 (2008).

40. Roman V et al., "Differential Effects of Chronic Partial Sleep Deprivation and Stress on Serotonin-1A and Muscarinic Acetylcholine Receptor Sensitivity." *J. Sleep Res.* 15, 386–94 (2006).

41. Bass J et al., "Sleepless in America: A Pathway to Obesity and the Metabolic Syndrome?" *Arch. Intern. Med.* 165, 15–6 (2005).

42. Greer SM et al., "The Impact of Sleep Deprivation on Food Desire in the Human Brain." *Nat. Commun.* 4, 2259 (2013).

43. Prather AA et al., "Short and Sweet: Associations Between Self-Reported Sleep Duration and Sugar-Sweetened Beverage Consumption Among Adults in the United States." *Sleep Health* 2, 272–76 (2016).

44. Perez-Cornago A et al., "Effect of Dietary Restriction on Peripheral Monoamines and Anxiety Symptoms in Obese Subjects with Metabolic Syndrome." *Psychoneuroendocrinology* 47, 98–106 (2014).

45. Fombonne E, "Increased Rates of Depression: Update of Epidemiological Findings and Analytical Problems." *Acta Psychiatr. Scand.* 90, 145–56 (1994).

CHAPTER 11. LIFE, LIBERTY, AND THE PURSUIT OF HAPPINESS?

1. Zitner A, "U.S. Seen on Wrong Track by Nearly Three-Quarters of Voters," *Wall Street Journal*, July 17, 2016. http://www.wsj.com/articles/u-s-seen-on-wrong-track-by-nearly-three-quarters-of-voters-1468760580

2. Crude and Age-Adjusted Death Rates for All Causes: 2014—Quarter 4, 2015. http://www.cdc.gov/nchs/products/vsrr/mortality-dashboard.htm

3. Mukherjee S, "U.S. Life Expectancy Just Dropped for the First Time in More Than Two Decades." *Fortune*, New York Dec. 8, 2016. http://fortune.com/2016/12/08/us-life-expectancy-drops/

4. Xu J et al., "Mortality in the United States, 2015." NCHS Data Brief, Centers for Disease Control and Prevention, Washington, D.C. (2016). https://www.cdc.gov/nchs/products/databriefs/db267.htm

5. Barnes N et al., "An Examination of Mindfulness-Based Programs in US Medical Schools." *Mindfulness*, doi: 10.1007/S12671-016-0623-8 (2016).

6. Arias E, "Changes in Life Expectancy by Race and Hispanic Origin in the United States, 2013–2014." National Center for Health Statistics, Centers for Disease Control and Prevention, Washington, D.C. (2016). http://www.cdc.gov/nchs/products/databriefs/db244.htm

7. Xu J et al., "Mortality in the United States, 2015." NCHS Data Brief, Centers for Disease Control and Prevention, Washington, D.C. (2016). https://www.cdc.gov/nchs/products /databriefs/db267.htm

8. Wang H et al., "Global, regional, and national levels of neonatal, infant, and under-5 mortality during 1990–2013: a systematic analysis for the Global Burden of Disease Study 2013." *Lancet* 384, 957-79 (2014).

9. Chetty R et al., "Where Is the Land of Opportunity? The Geography of Intergenerational Mobility in the United States." *Quart. J. Econ.* 129, 1553–623 (2014).

10. "Drug Overdose Deaths Reach All-Time High." CNN, Dec. 18, 2015. http://www.cnn .com/2015/12/18/health/drug-overdose-deaths-2014/index.html

11. Ingraham C, "Heroin Deaths Surpass Gun Homicides for the First Time, CDC Data Shows," *Washington Post*, Dec. 8, 2016. http://www.washingtonpost.com/news/wonk /wp/2016/12/08/heroin-deaths-surpass-gun-homicides-for-the-first-time-cdc-data -show/?utm_term=.247b47fc53d5

12. Case A et al., "Rising Morbidity and Mortality in Midlife Among White Non-Hispanic Americans in the 21st Century." *Proc. Natl. Acad. Sci.* 112, 15078–83 (2015).

13. Tavernise S, "White Americans Are Dying Younger as Drug and Alcohol Abuse Rise," *New York Times*, April 20, 2016. https://www.nytimes.com/2016/04/20/health/life -expectancy-decline-mortality.html?_r=0

14. Case A et al., "Mortality and Morbidity in the 21st Century." Brookings Papers on Economic Activity, March 23–24, 2017. https://www.brookings.edu/bpea-articles /mortality-and-morbidity-in-the-21st-century/

15. Lowry R et al., "Suicidal Thoughts and Attempts Among U.S. High School Students: Trends and Associated Health-Risk Behaviors, 1991–2011." *J. Adol. Health* 54, 100–108 (2014).

16. Xu J et al. "Mortality in the United States, 2015." NCHS Data Brief, Centers for Disease Control and Prevention, Washington, D.C. (2016). https://www.cdc.gov/nchs/products /databriefs/db267.htm

17. Samuelson RJ, "The Global Happiness Derby," *Washington Post*, April 15, 2012. http:// www.washingtonpost.com/opinions/the-global-happiness-derby/2012/04/15/giqajwi8jt _story.html

18. Wiencek H, *Master of the Mountain: Jefferson and His Slaves*. Farrar, Strauss, and Giroux, New York (2013).

19. Photo of George Mason Statue (2015). http://www.google.com/search?site=&tbm=isch &source=hp&biw=1400&bih=783&q=george+mason&oq=george+mason&gs_l=img.1.1.0l 10.410.4830.0.8477.18.13.0.0.0.0.997.1214.2-1j6-1.2.0....0...1.1.64.img..16.2.1212.hdk-3i _d8dw—imgrc=2iokiaotulvewm%3a

20. Choi HK et al., "Soft Drinks, Fructose Consumption, and the Risk of Gout in Men: Prospective Cohort Study." *BMJ* 336, 309–12 (2008).

21. Franklin B, "Dialog Between Franklin and the Gout," in *Oxford Book of American Essays*, 1914, Matthews B, ed. Bartleby, New York (2000).

22. "Key Finding: Prevalence of Disability and Disability Type Among Adultts, United States—2013." http://www.cdc.gov/ncbddd/disabilityandhealth/features/key-findings -community-prevalence.html

23. "Drug Overdose Deaths Reach All-Time High." CNN, Dec. 18, 2015. http://www.cnn .com/2015/12/18/health/drug-overdose-deaths-2014/index.html

24. Gaither JR et al., "National Rends in Hospitalizations for Opioid Poisonings Among Children and Adolescents, 1997 to 2012." *JAMA Pediatr.*, 170, 1195-201 (2016).
25. Wertheim LJ, "'Smack' Epidemic: How Painkillers Are Turning Young Athletes into Heroin Addicts.' *Sports Illustrated*, June, 22, 2015. http://www.si.com/vault/2016/02/11/smack-epidemic
26. Ibid.
27. *Saturday Night Live*, "Heroin AM," April 17, 2016. https://www.youtube.com/watch?v=W-ZdQ0z5cLM
28. Gounder C, "Who Is Responsible for the Pain Pill Epidemic?" *New Yorker*, Nov. 8, 2013. http://www.newyorker.com/business/currency/who-is-responsible-for-the-pain-pill-epidemic
29. Sontag D, "Addiction Treatment with a Dark Side." *New York Times*, Nov. 16, 2013. http://www.nytimes.com/2013/11/17/health/in-demand-in-clinics-and-on-the-street-bupe-can-be-savior-or-menace.html
30. Clement S et al., "One-Third of Long-Term Users Say They're Hooked on Prescription Opioids." *Washington Post*, Dec. 9, 2016. http://www.washingtonpost.com/national/health-science/one-third-of-long-term-users-say-theyre-hooked-on-prescription-opioids/2016/12/09/e048d322-baed-11e6-91ee-1adddfe36cbe_story.html?utm_term=.c37297ae5334
31. Newman R, "U.S. Incomes Are Finally Growing." http://finance.yahoo.com/news/u-s—incomes-are-finally-growing-161416587.html
32. Luthra S, "After Medical Marijuana Legalized, Medicare Presecriptions Drop for Many Drugs." http://www.npr.org/sections/health-shots/2016/07/06/484977159/after-medical-marijuana-legalized-medicare-prescriptions-drop-for-many-drugs
33. Crippa JA et al., "Cannabis and Anxiety: a Critical Review of the Evidence." *Human Psychopharmacol.* 24, 515–23 (2009).
34. Twomey CD, "Association of Cannabis Use with the Development of Elevated Anxiety Symptoms in the General Population: A Meta-Analysis." *J. Epidemiol. Commun. Health*, March 7, 2017, pii: Jech-2016-208145.
35. Kedzior KK et al., "A Positive Association Between Anxiety Disorders and Cannabis Use or Cannabis Use Disorders in the General Population—A Meta-Analysis of 31 Studies." *BMC Psychiatry* 14, 136 (2014).
36. Meier MH et al., "Persistent Cannabis Users Show Neuropsychological Decline from Childhood to Midlife." *Proc. Natl. Acad. Sci.* 109, E2657-E64 (2012).
37. Committee on the Health Effects of Marijuana: National Academies of Sciences, Engineering, and Medicine, *The Health Effects of Cannabis and Cannabinoids: The Current State of Evidence and Recommendations for Research.* National Academies Press, doi: 10.17226/24625 (2017).

CHAPTER 12. GROSS NATIONAL UNHAPPINESS

1. Bok D, *The Politics of Happiness: What Government Can Learn from the New Research on Well-Being.* Princeton University Press, Princeton (2010).
2. "Chanel and Burberry Drop Moss After Cocaine Claims," *Telegraph*, Sept. 21, 2005. http://www.telegraph.co.uk/news/1498858/chanel-and-burberry-drop-moss-after-cocaine-claims.html

3. Solnick S et al., "Is More Always Better? A Survey on Positional Concerns." *J. Econ. Behav. Org.* 37, 373–83 (1998).
4. Layard R, *Happiness: Lessons from a New Science*. Penguin, New York (2005).
5. Easterlin RA, "Does Economic Growth Improve the Human Lot? Some Empirical Evidence," in *Nations and Households in Economic Growth: Essays in Honor of Moses Abramovitz*, David PA and Reder MW, eds. Academic Press, New York (1974), 89–125.
6. Campanella E, "Is It Time to Abandon GDP?" Project Syndicate (2016). http://www .project-syndicate.org/onpoint/is-it-time-to-abandon-gdp?utm_source=op+newsletter +list&utm_campaign=d5fa489ac1-is_it_time_to_abandon_gdp_2016_11_04&utm _medium=email&utm_term=0_2514e9df8e-d5fa489ac1-105279581
7. Philipsen D, *The Little Big Number: How GDP Came to Rule the World and What to Do About It*. Princeton University Press, Princeton (2015).
8. The Prosperity Index. http://www.prosperity.com/—!/ranking
9. World Happiness Report. en.wikipedia.org/wiki/world_happiness_report
10. Happy Planet Index. en.wikipedia.org/wiki/happy_planet_index
11. Helman C, "The World's Happiest (And Saddest) Countries, 2013." *Forbes*, Oct. 29, 2013. http://www.forbes.com/sites/christopherhelman/2013/10/29/ the-worlds-happiest-and-saddest-countries-2013/—2ed328bd605526c48fde6055
12. Stevenson B et al., "Subjective Well-Being and Income: Is There Any Evidence of Satiation?" *Nat. Bureau Econ. Res.* (2013). http://www.brookings.edu/~/media/research /files/papers/2013/04/subjective well being income/subjective well being income.pdf
13. Kahneman D et al., "High Income Improves Evaluation of Life but Not Emotional Well-Being." *Proc. Natl. Acad. Sci.* 107, 16489–93 (2010).
14. Manuck SB et al. "Socio-Economic Status Covaries with Central Nervous System Serotonergic Responsivity as a Function of Allelic Variation in the Serotonin Transporter Gene-Linked Polymorphic Region." *Psychoneuroendocrinology* 29, 651–8 (2004).

CHAPTER 13. EXTREME MAKEOVER–WASHINGTON EDITION

1. Freudenberg N, *Lethal but Legal*. Oxford University Press, New York (2014).
2. Doll R et al., "Smoking and Carcinoma of the Lung." *Br. Med. J.* 2, 739–48 (1950).
3. Powell LJ, "Attack on the American Free Enterprise System" (1971). http:// reclaimdemocracy.org/powell_memo_lewis/
4. Stiglitz J, "The Secret Corporate Takeover." Project Syndicate (2016). http://www .project-syndicate.org/commentary/us-secret-corporate-takeover-by-joseph-e— stiglitz-2015-05
5. Federal Trade Commission, "In the Matter of Sugar Information, Inc. *et al.*" Federal Trade Commission, New York (1972), 711–24.
6. Scola N, "Exposing ALEC. How Conservative-Backed State Laws Are All Connected." *Atlantic*, April 14, 2012. http://www.theatlantic.com/politics/archive/2012/04/exposing -alec-how-conservative-backed-state-laws-are-all-connected/255869/
7. Wolin S, *Democracy Inc.: Managed Democracy and the Specter of Inverted Totalitarianism*. Princeton University Press, Princeton (2010).
8. Gilens M, *Affluence and Influence*. Princeton University Press, Princeton (2014).

CHAPTER 14. ARE YOU "LOVIN' IT"? OR "LIKING IT"?

1. Williams P, Happy (12AM). iamOTHER (2014). https://www.youtube.com/watch?v =i0A3-wc0rpw

2. Home S, "Intelligencer: Back and Forth." *New York*, March 1, 2009. http://nymag .com/news/intelligencer/55027/

3. Shield KD et al., "Alcohol Use and Breast Cancer: A Critical Review." *Alcohol Clin. Exp. Res.* 40, 1166–81 (2016).

4. Millward Brown, "Millward Brown Spreading Happiness with Coca-Cola" (2012). https://www.youtube.com/watch?v=deDRLmV-gpw

5. Eyal N, *Hooked: How to Build Habit-Forming Products*. Penguin, New York (2014).

6. Panova T et al. "Avoidance or Boredom: Negative Mental Health Outcomes Associated with Use of Information and Communication Technologies Depend on Users' Motivations." *Comp. Human Behav.* 58, 249–58 (2016).

7. Lepp A et al., "The Relationship Between Cell Phone Use and Academic Performance in a Sample of U.S. College Students." *SAGE Open* 5, Jan.–March 2015, 1–9, doi: 10.1177/2158244015573169.

8. Falbe J et al., "Sleep Duration, Restfulness, and Screens in the Sleep Environment." *Pediatrics* 135, E367–E375 (2015).

9. Sang-Hun C, "South Korea Expands Aid for Internet Addiction," *New York Times*, May 28, 2010. http://www.nytimes.com/2010/05/29/world/asia/29game.html?_r=0

10. Lin F et al., "Abnormal White Matter Integrity in Adolescents with Internet Addiction Disorder: A Tract-Based Spatial Statistics Study." *PLoS One* 7, e30253 (2012).

11. Zhu Y et al., "Molecular and Functional Imaging of Internet Addiction." *Biomed. Res. Int.* 10.1155/2015/378675, 2015:378675 (2015).

12. Guillot CR et al., "Longitudinal Associations Between Anhedonia and Internet-Related Addictive Behaviors in Emerging Adults." *Comput. Human Behav.* 62, 475–9 (2016).

13. Schou Andreassen C et al., "The Relationship Between Addictive Use of Social Media and Video Games and Symptoms of Psychiatric Disorders: A Large-Scale Cross-Sectional Study." *Psychol. Addict. Behav.* 30, 252–62 (2016).

14. Kuss DJ et al., "Internet Addiction: A Systematic Review of Epidemiological Research for the Last Decade." *Curr. Pharm. Des.* 20, 4026–52 (2014).

15. Selfhout MH et al., "Different Types of Internet Use, Depression, and Social Anxiety: The Role of Perceived Friendship Quality." *J. Adolesc.* 32, 819–33 (2009).

16. Tandoc E et al., "Facebook Use, Envy, and Depression Among College Students: Is Facebooking Depressing?" *Comp. Human Behav.* 43, 139–46 (2015).

17. Primack BA et al., "Association Between Media Use in Adolescence and Depression in Young Adulthood: ALongitudinal Study." *Arch. Gen. Psychiatry* 66, 181–8 (2009).

18. Zhu Y et al., "Molecular and Functional Imaging of Internet Addiction." *Biomed. Res. Int.* 10.1155/2015/378675, 2015:378675 (2015).

19. Sherman LE et al., "The Power of the Like in Adolescence: Effects of Peer Influence on Neural and Behavioral Responses to Social Media." *Psychol. Sci.* 27, 1027–35 (2016).

20. Coates J, *The Hour Between Dog and Wolf: How Risk Taking Transforms Us, Body and Mind*. Penguin, New York (2012).

21. Paramaguru K, "Coked-Up British Bankers Caused the Financial Crisis, Says UK Professor." *Time*, April 17, 2013. http://newsfeed.time.com/2013/04/17 /coked-up-british-bankers-caused-the-financial-crisis-says-uk-professor/

22. Friedman M, *Capitalism and Freedom*. University of Chicago Press, Chicago (1962).
23. Andreyeva T et al., "The Impact of Food Prices on Consumption: A Systematic Review of Research on the Price Elasticity of Demand for Food." *Am. J. Public Health* 100, 216–22 (2010).
24. Soto PL et al., "Dopamine D_2-Like Receptors and Behavioral Economics of Food Reinforcement." *Neuropsychopharmacology* 41, 971–8 (2016).
25. Centers for Medicare and Medicaid Services, "2015–2025 Projections of National Health Care Expenditures Data Released," July 13, 2016. http://www.cms.gov /newsroom/mediareleasedatabase/press-releases/2016-press-releases-items/2016 -07-13.html
26. Waterfield B, "Sugar Is 'Addictive and the Most Dangerous Drug of the Times,'" Telegraph, Sept. 17, 2013. http://www.telegraph.co.uk/news/worldnews/europe /netherlands/10314705/sugar-is-addictive-and-the-most-dangerous-drug-of-the -times.html

CHAPTER 15. THE DEATH SPIRAL

1. "First Future Health Index Research Indicates Americans Feel U.S. Healthcare Paradigm Needs Radical Shift over Next Decade." Philips Media, June 8, 2016. http:// www.usa.philips.com/a-w/about/news/archive/standard/news/press/2016/20160608 -first_future_health_index_research_indicates_americans_feel_us_healthcare _paradigm_needs_radical_shift_over_next_decade.html
2. Samocha-Bonet D et al., "Metabolically Healthy and Unhealthy Obese—The 2013 Stock Conference Report." *Obes. Rev.* 15, 697–708 (2014).
3. Voulgari C et al., "Increased Heart Failure Risk in Normal-Weight People with Metabolic Syndrome Compared with Metabolically Healthy Obese Individuals." *J. Am. Coll. Cardiol.* 58, 1343–50 (2011).
4. Abbasi F et al., "Relationship Between Obesity, Insulin Resistance, and Coronary Heart Disease Risk." *J. Am. Coll. Cardiol.* 40, 937–43 (2002).
5. Ford ES et al., "Prevalence of the Metabolic Syndrome Among US Adults: Findings from the Third National Health and Nutrition Examination Survey." *JAMA* 287, 356–9 (2002).
6. Xu J et al., *Mortality in the United States, 2015.* Centers for Disease Control and Prevention. Washington, D.C. (2016). https://www.cdc.gov/nchs/products/databriefs /db267.htm
7. Roth GA et al., "Demographic and Epidemiologic Drivers of Global Cardiovascular Mortality." *N. Engl. J. Med.* 372, 1333–41 (2015).
8. Chetty R et al., "The Association Between Income and Life Expectancy in the United States, 2001–2014." *JAMA* 315, 1750–66 (2016).
9. Case A et al., "Rising Morbidity and Mortality in Midlife Among White Non-Hispanic Americans in the 21st Century." *Proc. Natl. Acad. Sci.* 112, 15078–83 (2015).
10. "Global Burden of Metabolic Risk Factors for Chronic Diseases Collaboration, Cardiovascular Disease, Chronic Kidney Disease, and Diabetes Mortality Burden of Cardiometabolic Risk Factors from 1980 to 2010: A Comparative Risk Assessment." *Lancet Diabetes Endocrinol.* 2, 634–47 (2014).

11. Terry LL, *Smoking and Health: Report of the Advisory Committee to the Surgeon General of the Public Health Service*. U.S. Dept. of Health, Education, and Welfare, Washington, D.C. (1964). http://profiles.nlm.nih.gov/ps/access/nnbbmq.pdf

12. "Social Policy Lab: A Systemic Approach" (in Spanish). CISS, July 3, 2015. http://www .ciss.net/news_activity/social-policy-lab-a-systemic-approach/?lang=en

13. Ford ES et al., "Explaining the Decrease in U.S. Deaths from Coronary Disease, 1980–2000." *N. Engl. J. Med.* 356, 2388–98 (2007).

14. Danner D et al., "Positive Early Emotions in Life and Longevity: Findings from the Nun Study." *J. Personal. Soc. Psychol.* 80, 804–13 (2001).

15. Diener E et al., "Happy People Liver Longer: Subjective Well-Being Contributes to Health and Longevity." *App. Psychol. Health Well-Being* 3, 1–43 (2011).

16. Zaninotto P et al., "Sustained Enjoyment of Life and Mortality at Older Ages: Analysis of the English Longitudinal Study of Ageing." *BMJ* 355, doi: 10.1136/bmj .i6267 (2016).

17. Liu B et al., "Does Happiness Itself Directly Affect Mortality? The Prospective UK Million Women Study." *Lancet* 387, 874–81 (2016).

18. Tawakol A et al., "Relation Between Resting Amygdalar Activity and Cardiovascular Events: A Longitudinal and Cohort Study." *Lancet*, doi: 10.1016/S0140-6736(16)31714-7 (2017).

19. Emanuel E, "Why I Hope to Die at 75." *Atlantic Monthly*, Oct. 2014. http://www .theatlantic.com/magazine/archive/2014/10/why-i-hope-to-die-at-75/379329/

20. Olshansky SJ et al., "A Potential Decline in Life Expectancy in the United States in the 21st Century." *N. Engl. J. Med.* 352, 1138–45 (2005).

21. Himmelstein DU et al., "The Current and Projected Taxpayer Shares of US Health Costs." *Am. J. Public Health* 106, 449–52 (2016).

22. Roy A, "Obamacare's MLR 'Bomb' Will Create Private Insurance Monopolies and Drive Premiums Skyward. Hallelujah!" *Forbes*, Dec. 6, 2011. http://www.forbes.com /sites/theapothecary/2011/12/06/obamacares-mlr-bomb-will-create-private-insurance -monopolies-and-drive-premiums-skyward-hallelujah/—6044ee4c5acc

23. Mcgee S, "Obamacare Premiums and Deductibles Going Up—But It's Still Better Than Before," *Guardian*, Nov. 20, 2015. http://www.theguardian.com/money/us-money -blog/2015/nov/20/obamacare-premiums-deductibles-increase-health-care

24. Finkelstein EA et al., "The Costs of Obesity in the Workplace." *J. Occup. Environ. Med.* 52, 971–6 (2010).

25. Lustig RH, "Sickeningly Sweet: Does Sugar Cause Diabetes? Yes." *Can. J. Diabetes* 40, 282–7 (2016).

26. Singh GM et al., "Estimated Global, Regional, and National Disease Burdens Related to Sugar-Sweetened Beverage Consumption in 2010." *Circulation* 132, 639–66 (2015).

27. Food and Nutrition Service, *Foods Typically Purchased by Supplemental Nutrition Assistance Program (SNAP) Households (Summary)*. U.S. Dept. of Agriculture, Washington, D.C. (2016). http://www.fns.usda.gov/sites/default/files/ops /snapfoodstypicallypurchased-summary.pdf

28. Conrad Z et al., "Cardiometabolic Mortality by Supplemental Nutrition Assistance Program Participation and Eligibility in the United States." *Am. J. Public Health* 107, doi: 10.2105/ajph.2016.303608 (March 2017).

29. Vreman RA et al., "Non-Alcoholic Fatty Liver Disease as a Mediator of Detriments of Dietary Sugar Consumption: Implications for the Health and Economic Benefits of Interventions in the United States." *BMJ Open* (in press).

CHAPTER 16. CONNECT (RELIGION, SOCIAL SUPPORT, CONVERSATION)

1. Young SN, "How to Increase Serotonin in the Brain Without Drugs." *Rev. Psychiatr. Neurosci.* 32, 394–9 (2007).
2. Borg J et al., "Contribution of Non-Genetic Factors to Dopamine and Serotonin Receptor Availability in the Adult Human Brain." *Mol. Psychiatry* 21, 1077–84 (2016).
3. Reiss S, "The Sixteen Strivings for God." *Zygon* 39, 303–320, doi: 10.1111/j.1467 -9744.2004.00575.x (June 2004).
4. Diener E et al., "The Religion Paradox: If Religion Makes People Happy, Why Are So Many Dropping Out?" *J. Personal. Soc. Psychol.* 101, 1278–90 (2011).
5. Bullard G, "The World's Newest Major Religion: No Religion." *National Geographic*, April 2016. http://news.nationalgeographic.com/2016/04/160422-atheism-agnostic-secular-nones-rising-religion/
6. Greene RA, "Americans Not Losing Their Religion, but Changing It Often." CNN, April 27, 2009. http://cnn.com/2009/us/04/27/changing.religion.study/index.html
7. Bingham J, "Religion Can Make You Happier, Official Figures Suggest," *Telegraph*, Feb. 2, 2016. http://www.telegraph.co.uk/news/religion/12136531/religion-can-make -you-happier-official-figures-suggest.html
8. Lim C et al., "Religion, Social Networks, and Life Satisfaction." *Am. Sociol. Rev.* 75, 914–33 (2010).
9. Newport F et al., "In U.S., Very Religious Have Higher Wellbeing Across All Faiths." Gallup, Feb. 16, 2012. http://www.gallup.com/poll/152732/religious-higher-wellbeing -across-faiths.aspx
10. Hunter EV, "What Positive Psychologists and Mormons Can Learn from Each Other." University of Pennsylvania (2013). http://repository.upenn.edu/cgi/viewcontent.cgi ?article=1076&context=mapp_capstone
11. Diener E et al., "The Religion Paradox: If Religion Makes People Happy, Why Are So Many Dropping Out?" *J. Personal. Soc. Psychol.* 101, 1278–90 (2011).
12. Harris H et al., "The Neural Correlates of Religious and Nonreligious Belief." *PLoS One* 4, e0007272 (2009).
13. Borg J et al., "The Serotonin System and Spiritual Experiences." *Am. J. Psychiatry* 160, 1965–9 (2003).
14. Nilsson KW et al., "Genes Encoding for AP-2beta and the Serotonin Transporter Are Associated with the Personality Character Spiritual Acceptance." *Neurosci. Lett.* 411, 233–7 (2007).
15. Brewerton TD, "Hyperreligiosity in Psychotic Disorders." *J. Nerv. Ment. Dis.* 182, 302–4 (1994).
16. Macnamara P, "The God Effect." https://aeon.co/essays/the-dopamine-switch -between-atheist-believer-and-fanatic
17. Baumeister RF et al., "The Need to Belong: Desire for Interpersonal Attachments as a Fundamental Human Motivation." *Psychol. Bull.* 117, 497–529 (1995).

18. Southwick SM et al., "The Psychobiology of Depression and Resilience to Stress: Implications for Prevention and Treatment." *Ann. Rev. Clin. Psychol.* 1, 255–91 (2005).

19. Eisenberger NI et al., "Social Neuroscience and Health: Neurophysiological Mechanisms Linking Social Ties with Physical Health." *Nat. Neurosci.* 15, 669–74 (2012).

20. Ybarra O et al., "Mental Exercising Through Simple Socializing: Social Interaction Promotes General Cognitive Functioning." *Pers. Soc. Psychol. Bull.* 34, 248–53 (2008).

21. Perreau-Linck E et al., "In Vivo Measurements of Brain Trapping of A-[11C] Methyl-L-Tryptophan During Acute Changes in Mood States." *J. Psychiatry Neurosci.* 32, 430–34 (2007).

22. Brummett BH et al., "Perceived Social Support as a Predictor of Mortality in Coronary Patients: Effects of Smoking, Sedentary Behavior, and Depressive Symptoms." *Psychosom. Med.* 67, 40–5 (2005).

23. Cacioppo JT et al., "In the Eye of the Beholder: Individual Differences in Perceived Social Isolation Predict Regional Brain Activation to Social Stimuli." *J. Cogn. Neurosci.* 21, 83–92 (2009).

24. Light SN et al., "Empathy Is Associated with Dynamic Change in Prefrontal Brain Electrical Activity During Positive Emotion in Children." *Child Dev.* 80, 1210–31 (2009).

25. Goleman D et al., "Social Intelligence and the Biology of Leadership." *Harvard Bus. Rev.* 86, 74–81 (2008).

26. Fairhurst MT et al., "Being and Feeling in Sync with an Adaptive Virtual Partner: Brain Mechanisms Underlying Dynamic Cooperativity." *Cerebral Cortex* 23, 2592–600 (2013).

27. Koehne S et al., "Perceived Interpersonal Synchrony Increases Empathy: Insights from Autism Spectrum Disorder." *Cognition* 146, 8–15 (2016).

28. Diamond LM et al., "Emotion Regulation Across the Life Span: an Integrative Perspective Emphasizing Self-Regulation, Positive Affect, and Dyadic Processes." *Motiv. Emot.* 27, 125–56 (2003).

29. Christakis NA et al., "The Spread of Obesity in a Large Social Network over 32 Years." *N. Engl. J. Med.* 357, 370–9 (2007).

30. Fowler JH et al., "Dynamic Spread of Happiness in a Large Social Network: Longitudinal Analysis over 20 Years in the Framingham Heart Study." *BMJ* 337, A2338 (2008).

31. Kross E et al., "Social Rejection Shares Somatosensory Representations with Physical Pain." *Proc. Natl. Acad. Sci.* 108, 6270–75 (2011).

32. Valenzuela S et al., "Is There Social Capital in a Social Network Site? Facebook Use and College Students' Life Satisfaction, Trust, and Participation." *J. Comput. Mediat. Commun.* 14, 875–901 (2009).

33. Zhang C et al., "Longitudinal Psychosocial Factors Related to Symptoms of Internet Addiction Among Adults in Early Midlife." *Addict. Behav.* 62, 65–72 (2016).

34. Kramer ADI et al., "Experimental Evidence of Massive-Scale Emotional Contagion Through Social Networks." *Proc. Natl. Acad. Sci.* 111, 8788–90 (2014).

35. Coviello L et al., "Detecting Emotional Contagion in Massive Social Networks." *PLoS One* 9, e90315 (2014).

36. Park J et al., "When Perceptions Defy Reality: the Relationships Between Depression and Actual and Perceived Facebook Social Support." *J. Affect. Disord.* 200, 37–44 (2016).

37. Caplan SE, "Preference for Online Social Interaction: A Theory of Problematic Internet Use and Psychosocial Well-Being." *Commun. Res.* 30, 625–48 (2003).

38. Park J et al., "When Perceptions Defy Reality: the Relationships Between Depression and Actual and Perceived Facebook Social Support." *J. Affect. Disord.* 200, 37–44 (2016).

39. Kross E et al., "Facebook Use Predicts Declines in Subjective Well-Being in Young Adults." *PLoS One* 8, e69841 (2013).

40. Verduyn P et al., "Passive Facebook Usage Undermines Affective Well-Being: Experimental and Longitudinal Evidence." *J. Exp. Psychol. Gen.* 144, 480–88 (2015).

41. Chou HT et al., "'They Are Happier and Having Better Lives Than I Am': The Impact of Using Facebook on Perceptions of Others' Lives." *Cyberpsychol. Behav. Soc. Netw.* 15, 117–21 (2012).

42. Caplan SE, "Theory and Measurement of Generalized Problematic Internet Use: A Two-Step Approach." *Comp. Hum. Behav.* 26, 1089–97 (2010).

43. Ryan T et al., "The Uses and Abuses of Facebook: A Review of Facebook Addiction." *J. Behav. Addict.* 3.3, 133–48 (2014).

44. Yanes A, "The Unsocial Experiment: a Month Without Social Media." Huffington Post, May 25, 2016. http://www.huffingtonpost.com/arianna-yanes/a-month-without -social-media_b_10119504.html

45. "Louis C.K. Hates Cell Phones." *Late Night with Conan O'Brien.* TNT, Sept. 20, 2013. https://www.youtube.com/watch?v=5HbYScltf1c

CHAPTER 17. CONTRIBUTE (SELF-WORTH, ALTRUISM, VOLUNTEERISM, PHILANTHROPY)

1. Brickman P et al., "Lottery Winners and Accident Victims: Is Happiness Relative?" *J. Personal. Soc. Psychol.* 36, 917–27 (1978).

2. Nissle S et al., "Winning the Jackpot and Depression: Money Cannot Buy Happiness." *Int. J. Psych. Clin. Pract.* 6, 183–6 (2002).

3. Arvey RD et al., "Work Centrality and Post-Award Work Behavior of Lottery Winners." *J. Psychol.* 138, 404–20 (2004).

4. Economic Research Service, "U.S. Spending on Food Away from Home Higher Than on Food at Home in 2014." U.S. Dept. of Agriculture, Washington, D.C. (2016). https:// www.ers.usda.gov/data-products/chart-gallery/gallery/chart-detail/?chartId=78742

5. Barclay E, "Your Grandparents Spent More of Their Money on Food Than You Do." Salt, NPR, March 2, 2015. http://www.npr.org/sections/thesalt/2015/03/02/389578089 /your-grandparents-spent-more-of-their-money-on-food-than-you-do

6. Ball K, "Traversing Myths and Mountains: Addressing Socioeconomic Inequities in the Promotion of Nutrition and Physical Activity Behaviours." *Int. J. Behav. Nutr. Phys. Act.* 12, 142 (2015).

7. Kasser T, *The High Price of Materialism.* MIT Press, Cambridge (2002).

8. Kasser T et al., "The Relations of Maternal and Social Environments to Late Adolescents' Materialistic and Prosocial Aspirations." *Dev. Psychol.* 31, 907–14 (1995).

9. Kasser T et al., "A Dark Side of the American Dream: Correlates of Financial Success as a Life Aspiration." *J. Person. Soc. Psychol.* 65, 410–22 (1993).

10. Schmuck P et al., "Intrinsic and Extrinsic Goals: Their Structure and Relationship to Wellbeing in German and U.S. College Students." *Soc. Indic. Res.* 50, 225–41 (2000).

11. Kasser T et al., "Materialistic Values and Wellbeing in Business Students." *Eur. J. Soc. Psychol.* 32, 137–46 (2002).

12. Dittmar H et al., "The Relationship Between Materialism and Personal Well-Being: A Meta-Analysis." *J. Personal. Soc. Psychol.* 107, 879–924 (2013).

13. Martos T et al., "Life Goals and Well-Being: Does Financial Status Matter? Evidence from a Representative Hungarian Sample." *Soc. Indic. Res.* 105, 561–8 (2012).

14. Kasser T et al., "Changes in Materialism, Changes in Psychological Well-Being: Evidence from Three Longitudinal Studies and an Intervention Experiment." *Motiv. Emot.* 38, 1–22 (2014).

15. Lastovicka JL et al., "Truly, Madly, Deeply: Consumers in the Throes of Material Possession Love." *J. Consumer Res.* 38, 323–42 (2011).

16. Bauer MA et al., "Cuing Consumerism: Situational Materialism Undermines Personal and Social Well-Being." *Psychol. Sci.* 23, 517–23 (2012).

17. Kasser T et al., "Changes in Materialism, Changes in Psychological Well-Being: Evidence from Three Longitudinal Studies and an Intervention Experiment." *Motiv. Emot.* 38, 1–22 (2014).

18. Bailey C et al., "What Makes Work Meaningful—or Meaningless." MIT Sloan Management Review, Sept. 2016. http://sloanreview.mit.edu/article/what-makes-work-meaningful-or-meaningless/

19. Adams S, "Most Americans Are Unhappy at Work." *Forbes*, June 20, 2014. https://www.forbes.com/sites/susanadams/2014/06/20/most-americans-are-unhappy-at-work/#73499876341a

20. Koenigs M et al., "Irrational Economic Decision-Making After Ventromedial Prefrontal Damage: Evidence from the Ultimatum Game." *J. Neurosci.* 27, 951–6 (2007).

21. Zhu L et al., "Damage to Dorsolateral Prefrontal Cortex Affects Tradeoffs Between Honesty and Self-Interest." *Nat. Neurosci.* 17, 1319–21 (2014).

22. Crockett MJ et al., "Impulsive Choice and Altruistic Punishment Are Correlated and Increase in Tandem with Serotonin Depletion." *Emotion* 10, 855–62 (2010).

23. Aan Het Rot M et al., "Social Behaviour and Mood in Everyday Life: Effects of Tryptophan in Quarrelsome Individuals." *J. Psychiatry Neurosci.* 31, 253–62 (2006).

24. Crockett MJ et al., "Dissociable Effects of Serotonin and Dopamine on the Valuation of Harm in Moral Decision Making." *Curr. Biol.* 25, 1852–9 (2015).

25. Talhelm T et al., "Large-Scale Psychological Differences Within China Explained by Rice Versus Wheat Agriculture." *Science* 344, 603–8 (2014).

26. George DR et al., "Intergenerational Volunteering and Quality of Life for Persons with Mild to Moderate Dementia: Results from a 5-Month Intervention Study in the United States." *Am. J. Geriatr. Psych.* 19, 392–6 (2011).

27. Jenkinson CE et al., "Is Volunteering a Public Health Intervention? A Systematic Review and Meta-Analysis of the Health and Survival of Volunteers." *BMC Public Health* 13, 773 (2013).

28. Tabassum F et al., "Association of Volunteering with Mental Well-Being: A Lifecourse Analysis of a National Population-Based Longitudinal Study in the UK." *BMJ Open* 6, e011327, doi: 10.1136/BMJopen-2016-011327 (2016).

29. Schreier HM et al., "Effect of Volunteering on Risk Factors for Cardiovascular Disease in Adolescents: A Randomized Controlled Trial." *JAMA Pediatr.* 167, 327–32 (2013).

30. Dunn EW et al., "Spending Money on Others Promotes Happiness." *Science* 319, 1687–8 (2008).

31. Harbaugh WT et al., "Neural Responses to Taxation and Voluntary Giving Reveal Motives for Charitable Donations." *Science* 316, 1622–5 (2007).

32. Moll J et al., "Human Fronto-Mesolimbic Networks Guide Decisions About Charitable Donation." *Proc. Natl. Acad. Sci.* 103, 15623–8 (2006).

33. Steenbergen L et al., "Tryptophan Promotes Charitable Donating." *Front. Psychol.* 5, 1451 (2014).

CHAPTER 18. COPE (SLEEP, MINDFULNESS, EXERCISE)

1. Blanding M, "Workplace Stress Responsible for up to $190B in Annual U.S. Healthcare Costs." *Forbes*, Jan. 26, 2015. https://www.forbes.com/sites/hbsworkingknowledge/2015/01/26/workplace-stress-responsible-for-up-to-190-billion-in-annual-u-s-heathcare-costs/#65f40db6235a

2. Ferenczi EA et al., "Prefrontal Cortical Regulation of Brainwide Circuit Dynamics and Reward-Related Behavior." *Science* 351, aac9698 (2016).

3. Goldstein AN et al., "The Role of Sleep in Emotional Brain Function." *Ann. Rev. Clin. Psychol.* 10, 679–708 (2014).

4. Campos-Rodriguez F et al., "Continuous Positive Airway Pressure Improves Quality of Life in Women with OSA. A Randomized-Controlled Trial." *Am. J. Respir. Crit. Care Med.* 194, 1286–94 (2016).

5. Goel N et al., "Cognitive Workload and Sleep Restriction Interact to Influence Sleep Homeostatic Responses." *Sleep* 37, 1745–56 (2014).

6. Prather AA et al., "Behaviorally Assessed Sleep and Susceptibility to the Common Cold." *Sleep* 38, 1353–9 (2015).

7. Kessler RC et al., "Insomnia and the Performance of US Workers: Results from the America Insomnia Survey." *Sleep* 34, 1161–71 (2011).

8. Palmer CA et al., "Sleep and Emotion Regulation: An Organizing, Integrative Review." *Sleep Med. Rev.* 31, 6–16, doi: 10.1016/j.smrv.2015.12.006 (2016).

9. Kessler RC et al., "Insomnia and the Performance of US Workers: Results from the America Insomnia Survey." *Sleep* 34, 1161–71 (2011).

10. Morris DZ, "New French Law Bars Work Email After Hours." *Fortune*, Jan. 1, 2017. http://fortune.com/2017/01/01/french-right-to-disconnect-law/

11. Criss D, "Texas Day Care's Message to Parents: 'Get off Your Phone!'" CNN, Feb. 2, 2017. http://www.cnn.com/2017/02/02/us/phone-message-day-care-trnd/

12. Dinges DF et al., "The Benefits of a Nap During Prolonged Work and Wakefulness." *Work and Stress* 2, 138–53 (1988).

13. Kelley AM et al., "Cognition Enhancement by Modafinil: A Meta-Analysis." *Aviat. Space Environ. Med.* 83, 685–90 (2012).

14. Ferini-Strambi L et al., "Effects of Continuous Positive Airway Pressure on Cognitition and Neuroimaging Data in Sleep Apnea." *Int. J. Psychophysiol.* 89, 203–12 (2013).

15. Campos-Rodriguez F et al., "Continuous Positive Airway Pressure Improves Quality of Life in Women with OSA. A Randomized-Controlled Trial." *Am. J. Respir. Crit. Care Med.* 194, 1286–94 (2016).

16. Bei B et al., "Chronotype and Improved Sleep Efficiency Independently Predict Depressive Symptom Reduction After Group Cognitive Behavioral Therapy for Insomnia." *J. Clin. Sleep Med.* 11, 1021–7 (2015).

17. West KE et al., "Blue Light from Light-Emitting Diodes Elicits a Dose-Dependent Suppression of Melatonin in Humans." *J. Appl. Physiol.* 110, 619–26 (2011).

18. Wang J et al., "Perfusion Functional MRI Reveals Cerebral Blood Flow Pattern Under Psychological Stress." *Proc. Natl. Acad. Sci.* 102, 17804–9 (2005).

19. Shields GS et al., "The Effects of Acute Stress on Core Executive Functions: A Meta-Analysis and Comparison with Cortisol." *Neurosci. Biobehav. Rev.* 68, 651–68 (2016).

20. Rideout V et al., "Generation M2: Media in the Lives of 8- to 18-Year-Olds." Henry J. Kaiser Family Foundation (2010). https://kaiserfamilyfoundation.files.wordpress.com/2013/01/8010.pdf

21. Watson JM et al., "Supertaskers: Profiles in Extraordinary Multi-Tasking Ability." *Psychon. Bull. Rev.* 17, 479–85 (2010).

22. Medeiros-Ward N et al., "On Supertaskers and the Neural Basis of Efficient Multitasking." *Psychon. Bull. Rev.* 22, 876–83 (2015).

23. Ophir E et al., "Cognitive Control in Media Multitaskers." *Proc. Natl. Acad. Sci.* 106, 15583–7 (2009).

24. Loh KK et al., "Higher Media Multi-Tasking Activity Is Associated with Smaller Gray-Matter Density in the Anterior Cingulate Cortex." *PLoS One* 9, e106698 (2014).

25. Loh KK et al., "Higher Media Multi-Tasking Activity Is Associated with Smaller Gray-Matter Density in the Anterior Cingulate Cortex." *PLoS One* 9, e106698 (2014).

26. Becker MW et al., "Media Multitasking Is Associated with Symptoms of Depression and Social Anxiety." *Cyberpsychol. Behav. Soc. Netw.* 16, 132–5 (2013).

27. Chadick JZ et al., "Structural and Functional Differences in Medial Prefrontal Cortex Underlie Distractibility and Suppression Deficits in Ageing." *Nat. Commun.* 5, 4223 (2014).

28. Sullivan A, "I Used to Be a Human Being." *New York* magazine, New York (Sept. 18, 2016). http://nymag.com/selectall/2016/09/andrew-sullivan-technology-almost-killed-me.html

29. Weng HY et al., "Compassion Training Alters Altruism and Neural Responses to Suffering." *Psychol. Sci.* 24, 1171–80 (2013).

30. Kabat-Zinn J, *Full Catastrophe Living*. Bantam, New York (1990).

31. Gelles D, "Mediation in Real Life," *New York Times* (2016). http://www.nytimes.com/column/meditation-for-real-life

32. Tang YY et al., "The Neuroscience of Mindfulness Meditation." *Nat. Rev. Neurosci.* 16, 213–25 (2015).

33. Fox KC et al., "Is Meditation Associated with Altered Brain Structure? A Systematic Review and Meta-Analysis of Morphometric Neuroimaging in Meditation Practitioners." *Neurosci. Biobehav. Rev.* 43, 48–73 (2014).

34. Cole MA et al., "Simultaneous Treatment of Neurocognitive and Psychiatric Symptoms in Veterans with Post-Traumatic Stress Disorder and History of Mild Traumatic Brain

Injury: A Pilot Study of Mindfulness-Based Stress Reduction." *Mil. Med.* 180, 956–63 (2015).

35. Oishi K et al., "Critical Role of the Right Uncinate Fasciculus in Emotional Empathy." *Ann. Neurol.* 77, 68–74 (2015).

36. Daubenmier J et al., "Effects of a Mindfulness-Based Weight Loss Intervention in Adults with Obesity: A Randomized Clinical Trial." *Obesity* 24, 794–804 (2016).

37. Daubenmier J et al., "Mindfulness Intervention for Stress Eating to Reduce Cortisol and Abdominal Fat Among Overweight and Obese Women: An Exploratory Randomized Controlled Study." *J. Obes.* 10.1155/2011/651936, 651936 (2011).

38. Daubenmier J et al., "Mindfulness Intervention for Stress Eating to Reduce Cortisol and Abdominal Fat Among Overweight and Obese Women: An Exploratory Randomized Controlled Study." *J. Obes.* 10.1155/2011/651936, 651936 (2011).

39. Coryell WH et al., "Fat Distribution and Major Depressive Disorder in Late Adolescence." *J. Clin. Psychiatry* 77, 84–9 (2016).

40. Thakore J et al., "Increased Intraabdominal Fat in Major Depression." *Biol. Psychiatry* 41, 1140–2 (1997).

41. Daubenmier J et al., "Effects of a Mindfulness-Based Weight Loss Intervention in Adults with Obesity: A Randomized Clinical Trial." *Obesity* 24, 794–804 (2016).

42. Rottensteiner M et al., "Leisure-Time Physical Activity and Intra-Abdominal Fat in Young Adulthood: A Monozygotic Co-Twin Control Study." *Obesity* 24, 1185–91 (2016).

43. Goedecke JH et al., "The Effect of Exercise on Obesity, Body Fat Distribution and Risk for Type 2 Diabetes." *Med. Sport. Sci.* 60, 82–93 (2014).

44. Cooney GM et al., "Exercise for Depression." *Cochrane Database Syst. Rev.* 9, CD004366 (2013).

45. Santarelli L et al., "Requirement of Hippocampal Neurogenesis for the Behavioral Effects of Antidepressants." *Science* 301, 805–9 (2003).

46. Boecker H et al., "The Runner's High: Opioidergic Mechanisms in the Human Brain." *Cereb. Cortex* 18, 2523–31 (2008).

47. Raichlen DA et al., "Wired to Run: Exercise-Induced Endocannabinoid Signaling in Humans and Cursorial Mammals with Implications for the 'Runner's High.'" *J. Exp. Biol.* 215, 1331–6 (2012).

48. Fox KR, "The Influence of Physical Activity on Mental Well-Being." *Public Health Nutr.* 2, 411–8 (1999).

49. Helgadóttir B et al., "Training Fast or Slow? Exercise for Depression: a Randomized Controlled Trial." *Prev. Med.* 91, 123–31 (2016).

50. Carter T et al., "The Effect of Exercise on Depressive Symptoms in Adolescents: A Systematic Review and Meta-Analysis." *J. Am. Acad. Child Adolesc. Psychiatry* 55, 580–90 (2016).

51. Catalan-Matamoros D et al., "Exercise Improves Depressive Symptoms in Older Adults: An Umbrella Review of Systematic Reviews and Meta-Analyses." *Psychiatry Res.* 244, 202–9 (2016).

52. Peng YF et al., "Analyzing Personal Happiness from Global Survey and Weather Data: A Geospatial Approach." *PLoS One* 11, e0153638 (2016).

53. Alderman BL et al., "MAP Training: Combining Meditation and Aerobic Exercise Reduces Depression and Rumination While Enhancing Synchronized Brain Activity." *Trans. Psychiatry* 6, e726 (2016).

54. Dicarlo LA et al., "Patient-Centered Home Care Using Digital Medicine and Telemetric Data for Hypertension: Feasibility and Acceptability of Objective Ambulatory Assessment." *J. Clin. Hypertens* (Greenwich) 18, 901–6 (2016).

55. Rickard N et al., "Development of a Mobile Phone App to Support Self-Monitoring of Emotional Well-Being: A Mental Health Digital Innovation." *JMIR Ment. Health* 3, e49 (2016).

56. Edwards EA et al., "Gamification for Health Promotion: Systematic Review of Behaviour Change Techniques in Smartphone Apps." *BMJ Open* 6, e012447 (2016).

57. Schoeppe S et al., "Efficacy of Interventions That Use Apps to Improve Diet, Physical Activity and Sedentary Behaviour: A Systematic Review." *Int. J. Behav. Nutr. Phys. Act.* 13, 127 (2016).

58. Guertler D et al., "Engagement and Nonusage Attrition with a Free Physical Activity Promotion Program: The Case of 10,000 Steps Australia." *J. Med. Internet Res.* 17, E176 (2015).

59. Gualtieri L et al., "Can a Free Wearable Activity Tracker Change Behavior? The Impact of Trackers on Adults in a Physician-Led Wellness Group." *JMIR Res Protoc.* 5(4), Nov 30, 2016; E237.

CHAPTER 19. COOK (FOR YOURSELF, YOUR FRIENDS, YOUR FAMILY)

1. Lustig RH et al., "The Toxic Truth About Sugar." *Nature* 487, 27–9 (2012).

2. Johnson RK et al., "Dietary Sugars Intake and Cardiovascular Health. A Scientific Statement from the American Heart Association Circulation," Circulation 1201011–20 (2009).

3. Vos MB et al., "Added Sugars and Cardiovascular Disease Risk in Children: A Scientific Statement from the American Heart Association." *Circulation*, Aug. 22, 2016, pii: CIR.0000000000000439.

4. Ervin RB et al., "Consumption of Added Sugar Among U.S. Children and Adolescents, 2005–2008." National Center for Health Statistics, Centers for Disease Control and Prevention, NCHS Data Brief no. 87, Feb. 2012. http://www.cdc.gov/nchs/data /databriefs/db87.htm

5. Mager DR et al., "The Effect of a Low Fructose and Low Glycemic Index/Load (FRAGILE) Dietary Intervention on Indices of Liverfunction, Cardiometabolic Risk Factors, and Body Composition in Children and Adolescents with Nonalcoholic Fatty Liver Disease (NAFLD)." *J. Parenter. Enteral Nutr.* 39, 73–84 (2015).

6. Kalia HS et al., "The Prevalence and Pathobiology of Nonalcoholic Fatty Liver Disease in Patients of Different Races or Ethnicities." *Clin. Liver Dis.* 20, 215–24 (2016).

7. Carliner H et al., "Prevalence of Cardiovascular Risk Factors Among Racial and Ethnic Minorities with Schizophrenia Spectrum and Bipolar Disorders: A Critical Literature Review." *Compr. Psychiatry* 55, 233–47 (2014).

8. Imamura F et al., "Consumption of Sugar Sweetened Beverages, Artificially Sweetened Beverages, and Fruit Juice and Incidence of Type 2 Diabetes: Systematic Review, Meta-Analysis, and Estimation of Population Attributable Fraction." *BMJ* 351, h3576 doi: 10.1136/BMJ.H3576 (2015).

9. Calvo-Ochoa E et al., "Short-Term High-Fat-and-Fructose Feeding Produces Insulin Signaling Alterations Accompanied by Neurite and Synaptic Reduction and

Astroglial Activation in the Rat Hippocampus." *J. Cereb. Blood Flow Metab.* 34, 1001–8 (2014).

10. Kruger HS et al., "Neonatal Hippocampal Lesion Alters the Functional Maturation of the Prefrontal Cortex and the Early Cognitive Development in Pre-Juvenile Rats." *Neurobiol. Learn. Mem.* 97, 470–81 (2012).

11. Singh GM et al., "Estimated Global, Regional, and National Disease Burdens Related to Sugar-Sweetened Beverage Consumption in 2010." *Circulation* 132, 639–66 (2015).

12. Singh GM et al., "Estimated Global, Regional, and National Disease Burdens Related to Sugar-Sweetened Beverage Consumption in 2010." *Circulation* 132, 639–66 (2015).

13. Gangwisch JE et al., "High Glycemic Index Diet as a Risk Factor for Depression: Analyses from the Women's Health Initiative." *Am. J. Clin. Nutr.* 102, 454–63 (2015).

14. Lustig RH et al., "Isocaloric Fructose Restriction and Metabolic Improvement in Children with Obesity and Metabolic Syndrome." *Obesity* 24, 453–60 (2016).

15. Schwarz JM et al., "Impact of Dietary Fructose Restriction on Liver Fat, *De Novo* Lipogenesis and Insulin Kinetics in Children with Obesity." *Gastroenterology* (in press).

16. Gugliucci A et al., "Short-term Isocaloric Fructose Restriction Lowers ApoC-III Levels and Yields Less Atherogenic Lipoprotein Profiles in Children with Obesity and Metabolic Syndrome." *Atherosclerosis* 253, 171–7 (2016).

17. Suglia SF et al., "Soft Drinks Consumption Is Associated with Behavior Problems in 5-Year-Olds." *J. Pediatr.* 163, 1323–8 (2013).

18. Solnick SJ et al., "Soft Drinks, Aggression and Suicidal Behaviour in US High School Students." *Int. J. Inj. Contr. Saf. Promot.* 21, 266–73 (2014).

19. Shi Z et al., "Soft Drink Consumption and Mental Health Problems Among Adults in Australia." *Public Health Nutr.* 13, 1073–9 (2010).

20. Yu B et al., "Soft Drink Consumption Is Associated with Depressive Symptoms Among Adults in China." *J. Affect. Disord.* 172, 322–427 (2015).

21. Henriksen RE et al., "Loneliness, Social Integration and Consumption of Sugar-Containing Beverages: Testing the Social Baseline Theory." *PLoS One* 9, e104421 (2014).

22. Mietus-Snyder ML et al., "Childhood Obesity: Adrift in the 'Limbic Triangle.'" *Ann. Rev. Med.* 59, 119–34 (2008).

23. Epel ES et al., "The Reward-Based Eating Drive Scale: A Self-Report Index of Reward-Based Eating." *PLoS One* 9, e101350 (2014).

24. Mason AE et al., "Reduced Reward-Driven Eating Accounts for the Impact of a Mindfulness-Based Diet and Exercise Intervention on Weight Loss: Data from the SHINE Randomized Controlled Trial." *Appetite* 100, 86–93 (2016).

25. Wang GJ et al., "BMI Modulates Calorie-Dependent Dopamine Changes in Accumbens from Glucose Intake." *PLoS One* 9, e101585 (2014).

26. Filbey FM et al., "Reward Circuit Function in High BMI Individuals with Compulsive Overeating: Similarities with Addiction." *Neuroimage* 63, 1800–6 (2012).

27. Karlsson HK et al., "Obesity Is Associated with Decreased ß-Opioid but Unaltered Dopamine D2 Receptor Availability in the Brain." *J. Neurosci.* 35, 3959–65 (2015).

28. Tuominen L et al., "Aberrant Mesolimbic Dopamine-Opiate Interaction in Obesity." *Neuroimage* 122, 80–86 (2015).

29. Langleben DD et al., "Depot Naltrexone Decreases Rewarding Properties of Sugar in Patients with Opioid Dependence." *Psychopharmacology* 220, 559–64 (2012).

30. Mason AE et al., "Acute Responses to Opioidergic Blockade as a Biomarker of Hedonic Eating Among Obese Women Enrolled in a Mindfulness-Based Weight Loss Intervention Trial." *Appetite* 91, 311–20 (2015).

31. Mason AE et al., "Putting the Brakes on the 'Drive to Eat': Pilot Effects of Naltrexone and Reward-Based Eating on Food Cravings Among Obese Women." *Eat Behav.* 19, 53–6 (2015).

32. Daubenmier J et al., "Effects of a Mindfulness-Based Weight Loss Intervention in Adults with Obesity: A Randomized Clinical Trial." *Obesity* 24, 794–804 (2016).

33. Moss M, *Salt, Sugar, Fat: How the Food Giants Hooked Us.* Random House, New York (2013).

34. Garber AK et al., "Is Fast Food Addictive?" *Curr. Drug Abuse Rev.* 4, 146–62 (2011).

35. U.S. General Accounting Office, "Food Safety: FDA Should Strengthen Its Oversight of Food Ingredients Determined to Be Generally Recognized as Safe (GRAS)." US GAO, Washington, D.C. (2010). http://www.gao.gov/products/gao-10-246

36. Card MM et al., "Just a Spoonful of Sugar Will Land You Six Feet Underground: Should the Food and Drug Aministration Revoke Added Sugar's GRAS Status?" *Food and Drug Law Journal* 70, 395 (2015).

37. Lustig RH, "GRAS: Smoke It, Don't Eat It," Huffington Post, June 7, 2013. http://www.huffingtonpost.com/robert-lustig-md/fda-food-additives_b_3384629.html

38. Michail N, "Major Dutch Retailer to Cut Sugar Across Its Private Label Range." http://www.foodnavigator.com/market-trends/major-dutch-retailer-to-cut-sugar-across-its-private-label-range

39. Kaminska I, "Robert Lustig: Godfather of the Sugar Tax" *Financial Times*, London (2016). http://www.ft.com/cms/s/0/311a74ec-ed24-11e5-888e-2eadd5fbc4a4.html?siteedition=intl—axzz4hkmj2rq0

40. Popkin BM et al., "Sweetening of the Global Diet, Particularly Beverages: Patterns, Trends, and Policy Responses." *Lancet Diab. Endocrinol.* 4, 174–86 (2016).

41. Interview with Ted Miguel, economist at University of California, Berkeley. http://www.gatesfoundation.org/global-development/pages/ted-miguel-interview-water-sanitation-hygiene.aspx

42. Kearns CE et al., "Sugar Industry and Coronary Heart Disease Research: a Historical Analysis of Internal Industry Documents." *JAMA Intern. Med.* 176, 1680–5 (2016).

43. Kearns CE et al., "Sugar Industry Influence on the Scientific Agenda of the National Institute of Dental Research's 1971 National Caries Program: A Historical Analysis of Internal Documents." *PLoS Med.* 12, e1001798 (2015).

44. Bes-Rastrollo M et al., "Financial Conflicts of Interest and Reporting Bias Regarding the Association Between Sugar-Sweetened Beverages and Weight Gain: A Systematic Review of Systematic Reviews." *PLoS Med.* 10, e1001578 (2013).

45. O'Connor A, "Coca-Cola Funds Scientists Who Shift Blame for Obesity Away from Bad Diets," *New York Times*, Aug. 9, 2015. http://well.blogs.nytimes.com/2015/12/01/research-group-funded-by-coca-cola-to-disband/?_r=0

46. Aaron DG et al., "Supporting Public Health to Deflect Coke and Pepsi Sponsorship of National Health Organizations by Two Major Soda Companies." *Am. J. Prev. Med.* 52, 20–30 (2017).

47. Dewey C, "How Nutella Plans to 'Trick' You into Thinking It's Healthier Than It Is." *Washington Post* Wonkblog, Jan. 9, 2017. http://www.standard.net/National/2017/01/09/How-Nutella-plans-to-trick-you-into-thinking-its-healthier-than-it-is

48. Mandrioli D et al., "Relationship Between Research Outcomes and Risk of Bias, Study Sponsorship, and Author Financial Conflicts of Interest in Reviews of the Effects of Artificially Sweetened Beverages on Weight Outcomes: A Systematic Review of Reviews." *PLoS One* 11, e0162198 (2016).

49. Lustig RH, "Processed Food: An Experiment That Failed." *JAMA Pediatr.* 171, 212-4 (2017).

INDEX

Page numbers in *italics* indicate figures.